Deep Earth Electrical Conductivity

Edited by
Wallace H. Campbell

1990

Birkhäuser Verlag
Basel · Boston · Berlin

Reprint from Pure and Applied Geophysics
(PAGEOPH), Volume 134 (1990), No. 4

Editor's address:

Wallace H. Campbell
U.S. Geological Survey
Mailstop 968
Box 25046
Denver, CO 80225
USA

Library of Congress Cataloging-in-Publication Data

Deep earth electrical conductivity
  p. cm.
  ISBN 0-8176-2564-X
  1. Earth resistance.   2. Earth-Mantle.
QC809.E15D44   1990
551.1'16–dc20

Deutsche Bibliothek Cataloging-in-Publication Data

Deep earth electrical conductivity / ed. by Wallace H.
Campbell. – Reprint. – Basel ; Boston ; Berlin : Birkhäuser,
1990
  Aus: Pure and applied geophysics (PAGEOPH) ; Vol. 134 (1990)
  ISBN 978-3-0348-7437-3   ISBN 978-3-0348-7435-9 (eBook)
  DOI 10.1007/978-3-0348-7435-9
NE: Campbell, Wallace H. [Hrsg.]

Contents

509   Introduction to Deep Earth Electrical Conductivity, *W. H. Campbell*

511   Electromagnetic induction in the earth due to stationary and moving sources, *D. H. Boteler*

527   The new approach to global deep sounding, *E. B. Fainberg, A. V. Kuvshinov, L. P. Mishina and B. Sh. Singer*

533   The effect of the oceans and sedimentary cover on global magnetovariational field distribution, *A. V. Kuvshinov, O. V. Pankratov and B. Sh. Singer*

541   The magnetospheric disturbance ring current as a source for probing the deep earth electrical conductivity, *W. H. Campbell*

559   The conductosphere depth at equatorial latitudes as determined from geomagnetic daily variations, *S. Duhau and A. Favetto*

575   A deep geophysical study in the Baikal region, *A. M. Popov*

589   A new telluric KCl probe using Filloux's AgAgCl electrode, *A. Junge*

PAGEOPH, Vol. 134, No. 4 (1990)

0033–4553/90/040509–01$1.50 + 0.20/0

Introduction to Deep Earth Electrical Conductivity

The 6th Assembly of the International Association of Geomagnetism and Aeronomy was held at Exeter, England, from July 24 to August 4, 1989. Two scientific sessions at the meeting were concerned with determinations of electrical conductivity deep within the earth. Seven papers by researchers present at those sessions were collected for joint publication in this journal.

Although the general topic of earth mantle conductivity was more comprehensively covered in PAGEOPH, Vol. 125, Nos. 2/3 (1987), this new, but smaller, collection is representative of the continuing research directions since that earlier date. In the first paper, BOTELER considers the problem of induction interpretation for sources that are in motion, a common condition for the natural field variations. The next presentation by FAINBERG et al. introduces a novel method of geomagnetic sounding that employs the minimization of differences between the experimental and simulated field responses from earth conductivity profiles. KUVSHINOV et al. discuss the effects of surface inhomogeneities (from oceans and sedimentary cover) upon the global modeling of earth's deep conductivity. The subject of source field conditions is introduced by CAMPBELL who reviews the magnetospheric ring current properties that are important in deep conductivity studies.

The last three papers emphasize applications. DUHAU and FAVETTO evaluate the earth conductivity beneath the region of the equatorial electrojet. POPOV discusses the deep conductivity properties of the Baikal, Siberia rift zone. The final presentation by JUNG introduces a telluric probe design for overcoming stability problems often found in telluric probes used for periods longer than one day.

Because the earth's mantle conductivity appears to increase with depth from about 10^{-1} to 10^{-3} S/m at its boundary with the earth crust to values near 1 to 10 S/m near 600 to 1000 km depth, probing these depths requires strong source fields of long period that are of global scale. Such natural sources are quite variable, not easily modeled, and present unique problems to both the theoretician and experimentalist. Although the seven papers of the special issue of PAGEOPH address all these problems, the reader will clearly see why this interesting topic is still actively pursued by researchers.

W. H. CAMPBELL
13 September 1990

PAGEOPH, Vol. 134, No. 4 (1990)

0033–4553/90/040511–16$1.50 + 0.20/0

Electromagnetic Induction in the Earth due to Stationary and Moving Sources

D. H. BOTELER[1]

Abstract—A new approach to the theory of electromagnetic induction is developed that is applicable to moving as well as stationary sources. The source field is considered to be a standing wave generated by two waves travelling in opposite directions along the surface of the earth. For a stationary source the incident waves have velocities of the same magnitude, however for a moving source the velocities of the two incident waves are respectively increased and decreased by the velocity of the source. Electromagnetic induction in the earth is then considered as refraction of these waves and gives, for both stationary and moving sources, the magnetotelluric relation:

$$\frac{-E_y}{H_x} = \left(\frac{i\omega\mu}{\sigma}\right)^{1/2}\left(1 - i\frac{v^2}{\omega\mu\sigma}\right)^{-1/2}$$

where v is the wavenumber of the source, μ is the permeability ($4\pi \cdot 10^{-7}$) and σ is the conductivity of the earth. ω is the angular frequency of the variation observed on the earth. For a stationary source the observed frequency is the same as the source frequency, however the effect of moving a time-varying source is to make the observed frequency different from the frequency of the source. Failure to recognise this in previous studies led to some erroneous conclusions. This study shows that a moving source is *not* "electromagnetically broader" than a stationary source as had been suggested.

Key words: Electromagnetic induction, refraction, moving sources.

1. Introduction

The frequencies used for magnetotelluric studies range from a few minutes to many hours, and arise from a variety of sources. Rapid magnetic variations, known as pulsations, are produced by causes such as instabilities on the outer magnetosphere and fluctuations in ionospheric currents. In all cases the size of the source region is large compared to the distance moved by an observatory on the earth during one period of oscillation and so they can be regarded as purely temporal variations. In contrast, the regular daily magnetic variation seen at mid and low latitude stations is produced by the earth's rotation carrying the observatory past

[1] Research School of Earth Sciences and Physics Department, Victoria University of Wellington, P.O. Box 600, Wellington, New Zealand. Present address: Geophysics Division, Geological Survey of Canada, 1 Observatory Crescent, Ottawa, Ontario, K1A 0Y3, Canada.

the S_q current system which is fixed with respect to the magnetosphere on the sunward side of the earth. Seen from the earth this daily variation appears as a magnetic field propagating over the surface of the earth with a velocity equal to the speed of rotation of the earth (400 m/s at the equator). At high latitudes magnetic field variations are mostly due to the auroral electrojet and its associated field-aligned currents. The electrojet is known to extend up to 60° in longitude and produces a magnetic bay lasting between 15 minutes and 1 hour. During an hour a site will move 15° in longitude which is a significant fraction of the spatial extent of the source field. Also the latitude of the electrojet can change during the course of a disturbance. Consequently, the magnetic disturbances produced by the auroral electrojet are due to a combination of temporal and spatial variations. Thus the theory for electromagnetic induction in the earth needs to be applicable to both stationary and moving source fields.

The initial work on the magnetotelluric technique by CAGNIARD (1953), TIKHONOV (1950), and KATO and KIKUCHI (1950) assumed the source field to be an infinite plane wave vertically incident on the earth. WAIT (1954) and PRICE (1962) pointed out that the limited extent of the source fields could have a significant effect on the induced electric fields, but the situation examined by these authors was only for a stationary source varying in time. The opposite case of a source that is moving but constant in time was considered by TIKHONOV and LIPSKAYA (1952) and HUTTON (1969, 1972) who treated long period magnetic field variations as a wave which is propagated from east to west with the velocity of the earth's rotation. The general case of induction by fields produced by a source that is both moving and varying in time was addressed by HERMANCE (1978). However, there have been some errors in the interpretation of the results obtained by HERMANCE, so the theory for moving sources needs further examination.

In this paper I develop a new approach for examining electromagnetic induction in the earth that is applicable to both stationary and moving time-varying source fields. A stationary source field of limited spatial extent is treated as a standing wave produced by two waves travelling with the same speed in opposite directions along the surface of the earth. Electromagnetic induction in the earth is then considered as refraction of these waves, and an analytic expression is obtained for E/H of the refracted wave. A moving source can be included in the theory by vector addition of the source velocity to the surface wave velocities. It will be shown that a movement of the source "Doppler shifts" the observed frequency but the expression obtained for E/H is the same as for a stationary source.

2. Refraction of Surface Waves

First consider refraction of a single wave travelling along the surface of the earth. This corresponds to induction due to movement of the earth with respect to a constant current system such as produces the regular diurnal magnetic variation. Induction due to stationary and moving time-varying sources will subsequently be treated as refraction of combinations of such surface waves. For simplicity neglect the curvature of the earth and consider the wave as incident on the surface of a uniform half-space of conductivity σ. The coordinate system is shown in Figure 1, where x is northward, y eastward, and z is vertically down. The xz plane contains the incident and reflected/refracted waves, and the waves have a time dependence $e^{i\omega t}$.

For refraction at the surface the incident and refracted waves are related by Snell's law:

$$k_i \sin \theta_i = k_r \sin \theta_r \tag{1}$$

where k_i and k_r are the propagation constants of the incident and refracted waves respectively and θ_i and θ_r are the angles of incidence and refraction.

For an incident wave with perpendicular polarisation the electric field vector is parallel to the surface and is given by

$$E_y = Ae^{-ivx}e^{i\omega t} \tag{2}$$

where v is the source wavenumber. At the frequencies in which we are interested for magnetotelluric studies the wavelength in the z direction of the incident wave is the wavelength in free space and is enormously greater than the diameter of the earth. Therefore the phase variation between the ionosphere and the earth is negligible and planes of constant phase can be considered to be vertical. Thus equation (2) can be considered to represent a wave, with propogation constant k_i, given by

$$k_i = iv \tag{3}$$

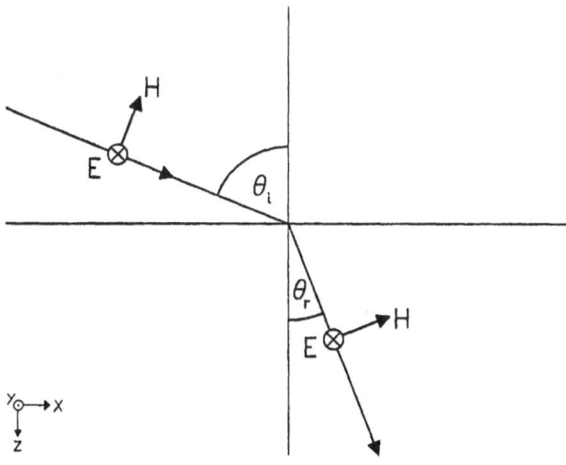

Figure 1
Refraction of a wave with the electric vector perpendicular to the plane of incidence (and so parallel to the surface).

and angle of incidence

$$\theta_i = 90°. \tag{4}$$

The wave with frequency ω is travelling in the positive x direction with velocity ω/v. In reality the waves are not freely propagating but are tied to the source. However, the velocity is that of planes of constant phase and so is still appropriately called the phase velocity.

The refracted wave has electric and magnetic field vectors related by the characteristic impedance of the medium

$$\frac{E}{H} = Z = \left(\frac{i\omega\mu}{\sigma}\right)^{1/2} \tag{5}$$

and propagates with the propagation constant and phase velocity that are also characteristic of the medium. When displacement currents are negligible this propagation constant is

$$k_r = (i\sigma\mu\omega)^{1/2}. \tag{6}$$

The angle of refraction can be found from Snell's law (equation 1) by substituting for k_i, k_r, and θ_i to give

$$\sin \theta_r = \frac{v^2}{i\sigma\mu\omega}. \tag{7}$$

Cos θ_r can be found from the relation

$$\cos \theta_r = (1 - \sin^2 \theta_r)^{1/2}. \tag{8}$$

Therefore

$$\cos \theta_r = \left(1 + \frac{v^2}{i\sigma\mu\omega}\right)^{1/2}. \tag{9}$$

For perpendicular incidence the refracted wave has an electric field vector parallel to the surface while the magnetic field vector is inclined to the surface. The horizontal electric and magnetic field components are therefore related by the expression

$$\frac{-E_y}{H_x} = \frac{E}{H}\frac{1}{\cos \theta_r}. \tag{10}$$

Substituting for E/H and cos θ_r then gives

$$\frac{-E_y}{H_x} = \left(\frac{i\omega\mu}{\sigma}\right)^{1/2}\left(1 - i\frac{v^2}{\omega\mu\sigma}\right)^{-1/2}. \tag{11}$$

A similar derivation can be done for parallel incidence and the expressions obtained in both cases are in agreement with those obtained by WAIT (1958).

However, only perpendicular incidence is applicable to magnetotelluric studies and this can now be considered as refraction of a travelling wave with a horizontal electric field and vertical magnetic field. This is consistent with PRICE's (1962) comment that in any physical situation the induced currents will flow in closed horizontal loops and will be determined by the electromotive forces generated by the changing vertical component of the inducing field.

3. Stationary Source

A stationary source produces horizontal and vertical magnetic field components with constituents of the form

$$H = Ae^{i\omega t}e^{vz} \cos vx \tag{12}$$

and

$$Z = -Ae^{i\omega t}e^{-vz} \sin vx. \tag{13}$$

Expanding the cos and sin terms gives

$$H = Ae^{-vz}\frac{e^{i(\omega t + vx)} + e^{i(\omega t - vx)}}{2} \tag{14}$$

and

$$Z = -Ae^{-vz}\frac{e^{i(\omega t + vx)} - e^{i(\omega t - vx)}}{2i} \tag{15}$$

showing that the magnetic field observed at the earth's surface can be synthesised from standing wave patterns generated by waves with space-time dependencies

$$e^{i(\omega t + vx)} \quad \text{and} \quad e^{i(\omega t - vx)}. \tag{16}$$

The waves have the same frequency but opposite propagation constant and phase velocity. Equation (16) can be considered to apply to two incident waves travelling northwards and southwards, respectively. In the northwards wave the magnetic field is upwards while in the southwards wave the magnetic field is downwards, and in both cases the electric field is westwards. Induction due to a stationary source can therefore be treated as refraction of identical waves travelling in opposite directions, as shown in Figure 2. The refracted waves produced both have northwards magnetic field components and westwards electric field components that are related by equation (11).

The vertical components of the refracted waves are equal and opposite, and thus cancel. Thus the resultant wave within the earth has planes of constant phase, as well as planes of constant amplitude, parallel to the surface. It can therefore be

STATIONARY SOURCE

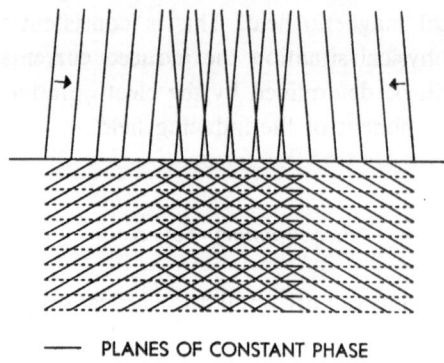

——— PLANES OF CONSTANT PHASE
····· PLANES OF CONSTANT AMPLITUDE

Figure 2
Refraction of the identical oppositely-directed travelling waves produced by a stationary source.

regarded as travelling vertically down with electric and magnetic fields related by the expression

$$\frac{-E_y}{H_x} = \left(\frac{i\omega\mu}{\sigma}\right)^{1/2}\left(1 - i\frac{v^2}{\omega\mu\sigma}\right)^{-1/2}.$$ (17)

This is consistent with the results obtained by PRICE (1962) and WAIT (1962).

4. Moving Source

HERMANCE (1978) has shown that a source moving with velocity v_x produces horizontal and vertical magnetic field components with constituents of the form

$$H = Ae^{i\omega_t t}\, e^{-vz} \cos(vx - vv_x t)$$ (18)

and

$$Z = -Ae^{i\omega_t t}e^{-vz} \sin(vx - vv_x t)$$ (19)

where ω_t is the frequency of the source.
Expanding the cos and sin terms now gives

$$H = Ae^{-vz}\frac{e^{i(\omega_t t + vx - vv_x t)} + e^{i(\omega_t t - vx + vv_x t)}}{2}$$ (20)

and

$$Z = -Ae^{-vz}\frac{e^{i(\omega_t t + vx - vv_x t)} - e^{i(\omega_t t - vx + vv_x t)}}{2i}.$$ (21)

Thus the magnetic field can be represented by a drifting, standing wave pattern generated by two travelling waves with velocities

$$v_+ = \frac{\omega_t}{v} + v_x \tag{22}$$

$$v_- = \frac{-\omega_t}{v} + v_x. \tag{23}$$

These waves have the space-time dependence

$$e^{i(\omega_t t + vx - vv_x t)} \quad \text{and} \quad e^{i(\omega_t t - vx + vv_x t)}. \tag{24}$$

which can be expressed as

$$e^{i[(\omega_t - vv_x)t + vx]} \quad \text{and} \quad e^{i[(\omega_t + vv_x)t - vx]}. \tag{25}$$

The magnetic field variations observed at the surface of the earth are due to a combination of temporal variations of the source and movement of spatial variations of the source relative to the earth. Temporal variations of the source can be considered to give rise to variations at the earth's surface with the "temporal" frequency, ω_t. The relative movement of spatial variations of the source will give rise to variations with a "spatial" frequency, ω_s, determined by the source wavenumber and drift velocity

$$\omega_s = vv_x. \tag{26}$$

The actual frequencies observed on the ground are the result of "beating" between the temporal and spatial frequencies and are given by

$$\omega_1 = \omega_t - vv_x \tag{27}$$

and

$$\omega_2 = \omega_t + vv_x. \tag{28}$$

Substituting into equation (25) then shows that the space-time dependence of the travelling waves can be written as

$$e^{i[\omega_1 t + vx]} \quad e^{i[\omega_2 t - vx]}. \tag{29}$$

Similarly using equation (26) to substitute for v_x in equations (22) and (23) gives

$$v_+ = \frac{\omega_t}{v} + \frac{\omega_s}{v} = \frac{\omega_2}{v} \tag{30}$$

$$v_- = \frac{\omega_t}{v} - \frac{\omega_s}{v} = \frac{-\omega_1}{v} \tag{31}$$

MOVING SOURCE

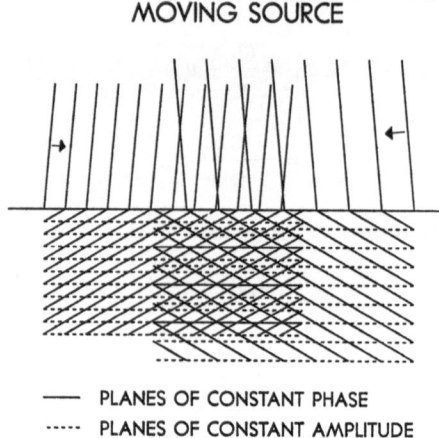

—— PLANES OF CONSTANT PHASE
····· PLANES OF CONSTANT AMPLITUDE

Figure 3

Refraction of the oppositely-directed travelling waves with different frequencies produced by a moving source.

showing that a moving source can be synthesised from two travelling waves with frequencies ω_1 and ω_2 and phase velocities $-\omega_1/v$ and ω_2/v.

Refraction of these waves at the surface produces waves that travel obliquely into the earth as shown in Figure 3. Both refracted waves have a vertical magnetic field component but because of the different frequencies of the waves, unlike the stationary source case, these vertical components do not cancel. The horizontal components of the electric and magnetic fields are related by the expressions

$$\frac{-E_y}{H_x} = \left(\frac{i\omega_1\mu}{\sigma}\right)^{1/2}\left(1 - i\frac{v^2}{\omega_1\mu\sigma}\right)^{-1/2} \tag{32}$$

$$\frac{-E_y}{H_x} = \left(\frac{i\omega_2\mu}{\sigma}\right)^{1/2}\left(1 - i\frac{v^2}{\omega_2\mu\sigma}\right)^{-1/2}. \tag{33}$$

In both these equations the relation is expressed in terms of the frequency observed at the ground and consequently is identical to the relation for a stationary source (equation 17).

5. Properties of the Refracted Waves

As the electromagnetic fields produced within the earth can now be considered as a refracted wave, considerable information can be obtained by the application of standard wave theory. (Caution is needed however as some well-known texts, e.g. STRATTON (1941), use a time dependence of $e^{-i\omega t}$ which reverses the sign of the imaginary component of the complex quantities derived.) For instance

$$\mathrm{Re}[k_r] = \text{constant} \tag{34}$$

defines planes of constant amplitude while

$$\mathrm{Imag}[k_r] = \text{constant} \tag{35}$$

defines planes of constant phases. The horizontal component of the propagation constant k_r is iv and has only an imaginary component whereas the vertical component

$$Q = (v^2 + i\omega\mu\sigma)^{1/2} \tag{36}$$

is complex. Thus planes of constant phase are inclined to the surface while planes of constant amplitude are parallel to the surface. When the planes of constant phase are inclined to the planes of constant amplitude the wave is said to have a complex angle of refraction, i.e. $\cos\theta_r$ is complex.

It will be seen that the vertical component, Q, of the propagation constant is the same as the induction parameter referred to in magnetotelluric theory. The real part of Q is the attenuation factor while the imaginary part is the phase constant. For

$$Q = a + ib \tag{37}$$

the depth at which the fields decay to $1/e$ of their surface value is given by

$$d = \frac{1}{a} \tag{38}$$

and the vertical component of the phase velocity within the earth is given by

$$v = \frac{\omega}{b}. \tag{39}$$

When $\omega\mu\sigma \gg v^2$ such that $\cos\theta_r = 1$ so that the individual refracted waves propagate vertically down

$$Q = (i\omega\mu\sigma)^{1/2} \tag{40}$$

and so

$$a = b = \left(\frac{\omega\mu\sigma}{2}\right)^{1/2} \tag{41}$$

therefore

$$d = \left(\frac{2}{\omega\mu\sigma}\right)^{1/2}. \tag{42}$$

This is dependent only on the properties of the medium and is the well-known expression for the skin depth. The phase velocity in this case is

$$v = \left(\frac{2\omega}{\mu\sigma}\right)^{1/2} \tag{43}$$

and again can be considered a property of the medium. The wavelength is given by

$$\lambda = \frac{\omega}{v} \tag{44}$$

and for this case

$$\lambda = d \tag{45}$$

where $\hat{\lambda} = \lambda/2\pi$, showing that the decay with depth is very rapid (the skin depth is less than one sixth of a wavelength), and currents only flow in one direction with no significant return currents at depth.

To determine the skin depth and vertical phase velocity when $\cos \theta_r \neq 1$ it is necessary to determine the real and imaginary parts of the induction parameter, Q, given by

$$Q^2 = (a + ib)^2 = v^2 + i\omega\mu\sigma. \tag{46}$$

Equating real and imaginary parts gives

$$a^2 - b^2 = v^2 \tag{47}$$

and

$$2ab = \omega\mu\sigma. \tag{48}$$

These equations have the solution

$$a^2 = \frac{v^2 \pm \sqrt{v^4 + (\omega\mu\sigma)^2}}{2} \tag{49}$$

and

$$b^2 = \frac{-v^2 \pm \sqrt{v^4 + (\omega\mu\sigma)^2}}{2} \tag{50}$$

where the sign of the root is determined by the need for a and b to be real. Thus the vertical phase velocity is

$$v = \left(\frac{2\omega}{\mu\sigma}\right)^{1/2} (\sqrt{1 + B^2} - B)^{-1/2} \tag{51}$$

where

$$B = \frac{v^2}{\omega\mu\sigma}. $$

Planes of constant phase now travel at an angle to the vertical and this gives rise to an increase in the vertical component of the phase velocity. The field now decays to $1/e$ of its surface value in a distance

$$p = \left(\frac{2}{\omega\mu\sigma}\right)^{1/2} (\sqrt{1 + B^2} + B)^{-1/2} \tag{52}$$

showing that the penetration of the field is dependent on the wavenumber of the source. Reducing the extent of the source (i.e., increasing the wavenumber) results in a greater decay with distance in air of the magnetic fields. This increased decay due to the source also contributes to the rate of decay of the magnetic fields within the conducting earth and therefore causes a reduced penetration depth. Induced currents tend to exclude the external magnetic field from the interior of a conducting medium and so a faster fall-off in the magnetic field due to a reduced source size means that lower current densities will achieve the same cancelling of the external field. Hence reducing the source size causes a decrease in the induced electric field and consequently a decrease in the E_y/H_x ratio as shown in equation (11).

6. Discussion

For source fields that are both moving and varying in time, special care needs to be taken when interpreting theoretical results. Magnetotelluric theory derives expressions in terms of the magnetic and electric fields observed at the surface of the earth. Thus ω is the angular frequency *observed*, not the angular frequency of the source. When the source is stationary the observed frequency is identical to the source frequency so the distinction is not obvious. However, when the source is moving the observed frequency and source frequency are not the same. There are two components: a "temporal" frequency, ω_t, due to time-variation of the source field; and a "spatial" frequency, $\omega_s = vv_x$, due to movement of the source field.

Calculation of the fields for realistic sources involves summation over a range of v values, as done by HERMANCE and PELTIER (1970). For example, the horizontal field for a line current, $I = I_0 e^{i\omega_t t}$, at a height h above the surface and horizontal distance x from the observer is

$$H = \frac{I_0}{2\pi} e^{i\omega_t t} \cdot \frac{h}{x^2 + h^2}. \tag{53}$$

Using the transform

$$\frac{h}{x^2 + h^2} = \int e^{-hv} \cos vx \, dv, \tag{54}$$

this can be written

$$H = \frac{I_0}{2\pi} e^{i\omega_t t} e^{-hv} \cos vx \, dv \tag{55}$$

and expanding the cosine term gives

$$H = \frac{I_0}{2\pi} e^{-hv} \frac{e^{i(\omega_t t + vx)} + e^{i(\omega_t t - vx)}}{2} \, dv. \tag{56}$$

Movement of the source changes the space-time dependence as shown in equation (25) and gives rise to a spatial frequency, ω_s, that is dependent on the source wavenumber. For a source that has a range of v values associated with it, any movement means that ω_s takes on a range of values. Thus the observed frequencies are

$$\omega = \omega_t \pm \omega_s(v) \quad \text{for} \quad v = 0 \to \infty \tag{57}$$

which represents the spectral broadening described by HERMANCE (1978).

HERMANCE (1978) also produced model results indicating that a moving source is "electromagnetically broader" than a stationary source, and CHAVE et al. (1981), using Hermance's theory, generated model results with a peak in apparent resistivity at a frequency dependent on the source velocity. These results, however, are incorrect and arise from confusing, observed frequency and source frequency.

The equations derived by HERMANCE (1978) give, for a uniform earth,

$$\frac{-E_y}{H_x} = \frac{i\mu(\omega_t \pm vv_x)}{Q} \tag{58}$$

where

$$Q^2 = v^2 + i\sigma\mu(\omega_t \pm vv_x)$$

which is correct, as can be seen if the substitution $\omega = \omega_t \pm v_x$ is made. In the case when the v^2 term is small, equation (54) becomes

$$\frac{-E_y}{H_x} = \left(\frac{i\mu}{\sigma}(\omega_t \pm vv_x)\right)^{1/2} \tag{59}$$

substituting $\omega_1 = \omega_t - vv_x$ and $\omega_2 = \omega_t + vv_x$ gives

$$\frac{-E_y}{H_x} = \left(\frac{i\omega_1\mu}{\sigma}\right)^{1/2} \quad \text{and} \quad \frac{-E_y}{H_x} = \left(\frac{i\omega_2\mu}{\sigma}\right)^{1/2}. \tag{60}$$

Averaging these impedance values produces an approximately correct result for the average frequency, ω_t, but with a bias towards the lower frequency value because of the square root involved in the expression.

However for modelling a uniform earth some authors have used the formula

$$\frac{-E_y}{H_x} = \frac{i\omega\mu}{Q} \tag{61}$$

with Q written as

$$Q = (v^2 + i\sigma\mu(\omega \pm vv_x))^{1/2} \tag{62}$$

to account for the movement of the source. This is incorrect because the ω in equation (61) is the frequency observed on the ground, while the ω in equation (62) is the frequency of the source; and, as explained, for a moving source these frequencies have different values. The effect of this mistake is to give a formula that, when the v^2 term is negligible, can be written

$$\frac{-E_y}{H_x} = \left(\frac{i\mu}{\sigma}\right)^{1/2} \left(\frac{\omega^2}{\omega \pm vv_x}\right)^{1/2}.$$

(63)

Evaluating this expression involves averaging the value obtained using $\omega + vv_x$ and the value obtained using $\omega - vv_x$. The effective frequency value obtained by this averaging is

$$\omega_e = \frac{1}{2}\left(\frac{\omega^2}{\omega + vv_x} + \frac{\omega^2}{\omega - vv_x}\right).$$

(64)

As vv_x increases, the second term starts to dominate and produces an effective frequency, ω_e, higher than the apparent value, ω. Thus producing impedance values for higher velocities is, in reality, producing values for higher frequencies. At higher frequencies the skin depth is reduced and so the ratio of source wavelength to skin depth is increased. The model therefore approaches more closely the conditions for the plane-wave approximation, thus giving the appearance that a moving source is "electromagnetically broader" than a stationary source.

Calculations made for increasing values of frequency, ω, would produce effective frequency values and impedance values that rise to a peak and then reduce. The frequency at which the peak occurs is given by $\omega = vv_x$ and so changes with velocity. Thus the peaks in apparent resistivity calculated by CHAVE et al. (1981) occur from ascribing the values to the wrong frequencies, not to any characteristics of moving sources.

Most sources such as the electrojets give rise to magnetic field variations which have a range of wavenumbers associated with each frequency component. In these situations, with present knowledge, the spatial extent is impossible to predict. However, for magnetic field variations due solely to rotation of the earth through spatial variations in the geomagnetic field, one component of the spatial extent is directly related to the observed frequency by the speed of rotation of the earth. In this case, substituting for v in equation (36) gives

$$Q^2 = v_x^2 + \frac{\omega^2}{v^2} + i\omega\mu\sigma$$

(65)

which is the expression used by HUTTON (1969, 1972). However, it should be noted that the velocity appears in this expression, not because of any special effect due to a moving source, but because it provides an easy means, in this case, of determining the source wavenumber at the observed frequencies.

It is often assumed in magnetotelluric work that the wave in the earth propagates vertically down, planes of constant phase are parallel to planes of constant amplitude, and the E and H vectors are parallel to the surface. Implicit in this approach is the assumption that the magnetic field variations are waves incident with a velocity, c, and the contrast in refractive index of the air and the earth is so great that for any angle of incidence the angle of refraction is expected to be zero. However, as has been shown, realistic sources produce fields that can be considered as waves travelling along the surface with a velocity dependent on the frequency and wavenumber of the source. This raises the possibility of the incident phase velocity being less than the phase velocity in the earth. In such a case an evanescent wave is produced which propagates parallel to the surface with the same velocity as the incident wave. Planes of constant amplitude are parallel to the surface while planes of constant phase are perpendicular to the surface.

The fundamental and harmonics of the quiet-day variation can be considered as a wave travelling at the speed of rotation of the earth which is 0.4 km/sec at the equator and 0.3 km/sec at mid-latitudes. Phase velocities in the earth, as a function of frequency, for conductivities ranging from that of seawater to that of high resistivity rock, are shown in Table 1. It will be seen that in medium and high conductivity areas the phase velocity in the earth is always less than the speed of rotation of the earth; so refraction at the surface will always occur. In areas of very low conductivity, the phase velocity in the earth for the fundamental is less than the incident phase velocity; so refraction again occurs. However, for the harmonics the phase velocity

Table 1

Phase velocities in the earth (km/sec). (A period of 24 hours corresponds to a frequency of 0.0000116 Hz.)

Frequency (Hz)	Conductivity (S/m)		
	3	10^{-1}	10^{-3}
1	1.8	10	100
0.1	0.57	3	30
0.01	0.18	1	10
0.001	0.057	0.3	3
0.0001	0.018	0.1	1
0.00001	0.006	0.03	0.3

in very low conductivity regions is greater than the incident phase velocity and so an evanescent wave would be expected to occur. In such conditions the relation between the electric and magnetic field vectors is dominated by the source characteristics and thus cannot be used to obtain reliable estimates of the earth's conductivity.

7. Conclusions

Electric fields are induced in the earth by a variation in the magnetic field observed at (and below) the surface of the earth. These observed magnetic field variations can be produced by actual temporal variations of an ionospheric or magnetospheric current system; or by movement of the earth relative to such a current system, either due to rotation of the earth or due to movement of the current system; or by a combination of both temporal and spatial variations. A stationary source can be considered as a standing wave pattern generated by two waves travelling with the same speed in opposite directions along the surface of the earth; and a moving source can be considered as a drifting standing wave pattern by vector addition of the source velocity to the surface wave velocities.

Electromagnetic induction in the earth can be viewed as refraction of an electromagnetic wave provided it is recognised that the incident wave is not propagating vertically down with the velocity of light but is travelling along the surface of the earth with a velocity dependent on the observed frequency and the horizontal wavenumber of the source. This allows derivation of the magnetotelluric relation for a stationary or moving source:

$$\frac{-E_y}{H_x} = \left(\frac{i\omega\mu}{\sigma}\right)^{1/2}\left(1 - i\frac{v^2}{\omega\mu\sigma}\right)^{-1/2}. \tag{66}$$

It is important to note that ω is the frequency observed on the ground and that, for a moving source, this is not the same as the source frequency.

The new theoretical approach agrees with the results obtained for a stationary source by PRICE (1962) and WAIT (1962) and for a constant moving source by HUTTON (1969, 1970). However, the present work shows that, in the interpretation of HERMANCE's (1978) results for a moving time-varying source, there was confusion between source frequency and observed frequency that led to some erroneous conclusions. The results obtained show that a moving source is *not* "electromagnetically broader" than a stationary source as suggested by HERMANCE; and neither can it give rise to the peak in the apparent resistivity reported by CHAVE *et al.* (1981).

Acknowledgement

I am particularly grateful to M. R. Ingham for his detailed comments on the manuscript.

REFERENCES

CAGNIARD, L. (1953), *Basic Theory of the Magneto-telluric Method of Geophysical Prospecting*, Geophysics *18*, 605–635.

CHAVE, A. D., VON HERZEN, R. P., POEHLS, K. A., and COX, C. S. (1981), *Electromagnetic Induction Fields in the Deep Ocean North-east of Hawaii: Implications for Mantle Conductivity and Source Fields*, Geophys. J. R. Astr. Soc. *66*, 379–406.

HERMANCE, J. F. (1978), *Electromagnetic Induction in the Earth by Moving Ionospheric Current Systems*, Geophys. J. R. Astr. Soc. *55*, 557–576.

HERMANCE, J. F., and PELTIER, W. R. (1970), *Magnetotelluric Fields of a Line Current*, J. Geophys. Res. *75*, 3351–3356.

HUTTON, R. (1969), *Electromagnetic Induction in the Earth by the Equatorial Electrojet*, Nature *222*, 363–364.

HUTTON, R. (1972), *Some Problems of Electromagnetic Induction in the Equatorial Electrojet Region — I. Magneto-telluric Relations*, Geophys. J. R. Astr. Soc. *28*, 267–284.

KATO, Y., and KIKUCHI, T. (1950), *On the Phase Difference of Earth Current Induced by the Changes of the Earth's Magnetic Field*, Sci. Rep. Tohoku. Univ. Ser. V. Geophysics *2*, 139–145.

PRICE, A. T. (1962), *The Theory of Magnetotelluric Methods when the Source Field is Considered*, J. Geophys. Res. *67*, 1907–1918.

STRATTON, J. A., *Electromagnetic Theory* (McGraw-Hill, New York 1941) pp. 15.

TIKHONOV, A. N. (1950), *Determination of the Electrical Characteristics of the Deep Strata of the Earth's Crust*, Dok. Akad. Nauk, USSR *73* (2), 295–297.

TIKHONOV, A. N., and LIPSKAYA, N. V. (1952), *Terrestrial Electric Field Variations*, Dok. Akad. Nauk. *87*, 547–550.

WAIT, J. R. (1954), *On the Relation between Telluric Currents and the Earth's Magnetic Field*, Geophysics *19*, 281–289.

WAIT, J. R. (1958), *Transmission and Reflection of Electromagnetic Waves in the Presence of Stratified Media*, J. Res. Nat. Bur. Stand. *61*, 505–232.

WAIT, J. R. (1962), *Theory of Magneto-telluric Fields*, J. Res. Nat. Bur. Stand. *66D*, 509–541.

(Received February 7, 1990, accepted April 4, 1990)

PAGEOPH, Vol. 134, No. 4 (1990)

0033–4553/90/040527–05$1.50 + 0.20/0

The New Approach to Global Deep Sounding

E. B. Fainberg,[1] A. V. Kuvshinov,[1] L. P. Mishina[1] and B. Sh. Singer[1]

Abstract —The new approach to global geomagnetic sounding is developed to overcome difficulties of spherical harmonical analysis and subsequent transfer function determination. The approach is based on minimizing the discrepancy between experimental and simulated magnetic fields. The discrepancy is considered as a function of the medium model parameters and the coefficients of external fields. The method can be used for laterally inhomogeneous as well as homogeneous earth models. An example of its application to a radially symmetric model is demonstrated.

Key words: Global deep sounding, nonuniform earth, numerical modelling.

1. Introduction

Conventional methods of global deep sounding are based on the one-dimensional earth model, whose conductivity distribution $\sigma(r)$ depends only upon the distance from the earth's center. These methods employ a spherical harmonical analysis procedure to separate the internal and external parts of a geomagnetic field and to evaluate transfer functions for a set of periods. Subsequently the conductivity distribution can be determined within a specified class of models by fitting the simulated transfer functions to the experimental one.

A serious difficulty occurs this way. It is connected with the known fact that the processing results obtained by different authors appear to be considerably different. Certainly, the observed scattering can be partially explained by the effects of inhomogeneities, but there exists another source of the scattering, internally connected with the spherical harmonical analysis. The most straightforward way to calculate spherical harmonical coefficients is to interpolate the magnetic field observed in a discrete set of observatories along the earth's surface and then to integrate them with the spherical functions. The difficulty is that magnetic observatories are extremely nonuniformly distributed. The largest part of the observatories is located in the Northern Hemisphere, mainly in Europe, and only a few of them

[1] Institute of Terrestrial Magnetism, Ionosphere and Radiowave Propagation of the USSR Academy of Sciences (IZMIRAN), 142092 Troitsk, Moscow Region, USSR.

are located in the Southern Hemisphere, in Africa and Asia. The oceans occupying 2/3 of the earth's surface are practically free of magnetic observatories. This makes the procedure of interpolation and the subsequent transfer function evaluation very ambiguous. The spherical harmonical coefficients could also be calculated with the assistance of the least-squares method, but the method can falter even on more serious difficulties of the same nature as the straightforward integration (FAINBERG, 1983a).

The progress of deep soundings is undoubtedly connected with numerical modelling of electromagnetic fields in the laterally nonuniform earth. Now it is possible, for example, to calculate fields excited by an arbitrary source in the model, consisting of a realistic surface layer and a laterally uniform medium below it. As the conductance of the surface layer is well-known one can find the conductivity distribution $\sigma(r)$ of the underlying medium fitting the simulated fields to the observed one. In this case the question arises for which external field the induced field should be calculated. Either the spherical harmonical analysis or method of surface integrals (PRICE and WILKINS, 1963) is used to separate the external and internal fields, the same problem of integration on a poor and nonuniform mesh exists.

To avoid unfounded assumptions on the fields behavior we shall try to develop a new method of analysis and interpretation. The initial idea of the method is to solve the inverse problem not only with respect to the medium parameters but to external fields parameters also.

2. Description of the Method

Let us consider a class of models being distinguished by a set of parameters $\{\sigma\}$. The class can, for example, consist of models with the same inhomogeneous surface layer and different underlying laterally uniform media. In this case parameters $\{\sigma\}$ characterize the underlying structure. We assume that for any model of the class the Green tensor function G_ω can be calculated. Therefore if $\mathbf{j}_\omega^e$ is the external source current density at frequency ω then the magnetic field in point $\mathbf{r}_i$ (the i-th observatory) is

$$G_\omega(\mathbf{r}_i, \{\sigma\}) * \mathbf{j}_\omega^e, \tag{1}$$

where symbol "$*$" denotes the convolution operation. All observations available at frequency ω can then be represented by a vector $\mathbf{H}_\omega = [\mathbf{H}_\omega(\mathbf{r}_i)]_{i=1,2,\dots}^T$ in a finite-dimensional Hilbert space (every frequency involved in analysis corresponds to its own space).

We assume also that the external source current is composed of a number of principal harmonics $\{\mathbf{f}_\nu\}_{\nu=1,2,\dots}$

$$\mathbf{j}_\omega^e = \sum_\nu a_\omega^\nu \mathbf{f}_\nu. \tag{2}$$

The total set of basic vectors is proposed to be the same at all frequencies, but only a few particular basic vectors are included in the external field of the particular frequency ω. The spherical harmonical functions can be used, for example, to compose this set. The observation vector can in this case be expressed in a form

$$\mathbf{H}_\omega = \gamma_\omega(\{\sigma\}) \cdot \mathbf{A}_\omega, \tag{3}$$

where matrix $\gamma_\omega(\{\sigma\}) = [G_\omega(\mathbf{r}_i, \{\sigma\}) * f_j]_{i=1,2,\ldots}^{j=1,2,\ldots}$ and vector $\mathbf{A}_\omega = [a_\omega^v]_{v=1,2,\ldots}^T$ and superscript "T" denotes the transposition operation.

We want to find the medium parameters together with the external fields parameters to satisfy the experimental data as well as possible. Hence we should satisfy the condition

$$\min_{\{\sigma\},\{\mathbf{A}_\omega\}} \sum_\omega p_\omega \|\mathbf{H}_\omega - \gamma_\omega(\{\sigma\}) \cdot \mathbf{A}_\omega\|_\omega^2, \tag{4}$$

where $\{p_\omega\}$ are weight coefficients and minimization is carried out with respect to both the medium parameters $\{\sigma\}$ and the external field coefficients $\{\mathbf{A}_\omega\}$. The denotation $\|\cdot\|_\omega$ is used for a norm into Hilbelt space corresponding to frequency ω. Note, that the major part of variables (i.e., coefficients $\{\mathbf{A}_\omega\}$) is involved linearly in the objective function, while the medium parameters appear in condition (4) linearly. A number of papers (LAWSON and SYLVESTRE, 1971; GOLUB and PEREYRA, 1973; RUHE and WEDIN, 1980) is devoted to the construction of algorithms to solve nonlinear separable least-squares problems. Below we shall describe one of the approaches.

For fixed $\{\sigma\}$ the minimizing condition with respect to A_ω is

$$\mathbf{A}_\omega = \beta_\omega(\{\sigma\})\mathbf{H}_\omega, \tag{5}$$

where

$$\beta_\omega(\{\sigma\}) = [\tilde{\gamma}_\omega^T(\{\sigma\})\gamma_\omega(\{\sigma\})]^{-1}\tilde{\gamma}_\omega^T(\{\sigma\}). \tag{6}$$

The necessary condition with respect to the $\{\sigma\}$ has the form

$$\min_{\{\sigma\}} \sum_\omega p_\omega \|\mathbf{H}_\omega - \gamma_\omega(\{\sigma\}) \cdot \mathbf{A}_\omega\|_\omega^2. \tag{7}$$

Remember that we have one system (5) for each frequency and all of them are interacting due to condition (7). Thus to solve the problem we should satisfy a huge system of conditions (5) and (7). In the case when the model is known we can find the coefficient vector $\mathbf{A}_\omega$ at frequency ω independently of other frequencies. On the other hand, if the external field for all frequencies is known the problem reduces to the usual inverse problem (7) to be solved only with respect to the medium parameters.

Conditions (5) and (7) should be satisfied simultaneously in the point $\{\sigma\}$, $\{\mathbf{A}_\omega\}$ of the parameters' space where functional $\sum_\omega p_\omega \|\mathbf{H}_\omega - \gamma_\omega(\{\sigma\}) \cdot \mathbf{A}_\omega\|_\omega^2$ attains its minimum value. The set of equations (5) determines some surface in this space. Therefore, instead of searching the minimum point along the whole space we can restrict ourselves to this surface. The objective function is equal then to $\Sigma_\omega p_\omega \|\mathbf{H}_\omega - \gamma_\omega(\{\sigma\})\beta_\omega(\{\sigma\}) \cdot \mathbf{H}_\omega\|_\omega^2$ and thus depends only on the medium parameters $\{\sigma\}$. Thus the problem (4) reduces to the condition

$$\min_{\{\sigma\}} \sum_\omega p_\omega \|\mathbf{H}_\omega - \gamma_\omega(\{\sigma\})\beta_\omega(\{\sigma\}) \cdot \mathbf{H}_\omega\|_\omega^2. \tag{8}$$

The functional now has the same number of arguments as it would have if we knew the external field explicitly. The single point which makes calculations in our case more extensive is that it is necessary to calculate the model response for all external field harmonics involved in the analysis in order to construct matrices $\gamma_\omega(\{\sigma\})$.

3. The Numerical Example

To test the method we synthesized the magnetic field for a laterally uniform model with conductivity distribution

$$\sigma(r) = \sigma_T \qquad\qquad r_0 - h_T < r < r_0$$

$$= \sigma_0 \cdot \left(\frac{r_0 - r}{h_0}\right)^\gamma \quad 0 < r < r_0 - h_T \tag{9}$$

where $\sigma_T = h_T/T = \sigma_0(h_T/h_0)^\gamma$, $h_0 = 400$ km, r_0 is the earth's radius, $T = 3 \cdot 10^8$ Ohm $\cdot$ m^2, $\sigma_0 = 1.5 \cdot 10^{-2}$ S/m, $\gamma = 4.735$ in accordance with (FAINBERG, 1983b) and (FAINBERG *et al.*, 1990). These parameters agree with experimental data received from global geomagnetic sounding. The magnetic field was synthesized at 18 points of the earth's surface. Zonal harmonics P_1^0, P_3^0 and first sectorial P_1^1 in periods ranging from 3 days until 6 months were included in the external field spectrum. To make data more realistic 10% noise was added to the synthesized field. The inversion was carried out with respect to the medium parameters σ_0, γ and the external field coefficients. The resultant values practically coincided with the initial ones.

The method was also applied to an analysis of experimental data from 19 magnetic observatories for the year 1965. After the band-pass filtration the data in the range of periods from 4 to 32 days were used for σ_0 and γ determination. Spherical harmonics $P_1^0, P_1^1, P_2^1, P_3^0$ were included in the external field spectrum. Derived values are close to those found in FAINBERG (1983b) on the basis of conventional technique.

In conclusion, it should be noted that the proposed approach can be used both for global sounding of the earth by variations of external and internal origins and for continental and regional sounding. The results of its application also can be useful for the reconstruction of ionospheric and magnetospheric currents being responsible for the magnetic field variations.

REFERENCES

GOLUB, G. H., and PEREYRA, V. (1973), *The Differentiation of Pseudo-inverses and Nonlinear Least-squares Problems Whose Variables Separate*, SIAM J. Numer. Anal. *10*, 413–432.

FAINBERG, E. B., *Global and Regional Magnetovariational Sounding of the Earth*, Ph.D. Thesis (IZMIRAN, Moscow 1983a) 430 pp. (in Russian).

FAINBERG, E. B., *Global geomagnetic sounding*, In *Mathematical Modeling of Electromagnetic Fields* (IZMIRAN, Moscow 1983b), pp. 79–121 (in Russian).

FAINBERG, E. B., KUVSHINOV, A. V., and SINGER, B. Sh. (1990), *Electromagnetic Induction in a Spherical Earth with Nonuniform Oceans and Continents in Electric Contact with the Underlying Medium: II—Bimodal Global Geomagnetic Sounding of the Lithosphere*, Geophys. J. (in press).

LAWTON, W. H., and SYLVESTRE, E. A. (1971), *Elimination of Linear Parameters in Nonlinear Regression*, Technometrics *13*, 4–13.

PRICE, A. T., and WILKINS, G. A. (1963), *New Methods for Analysis of Geomagnetic Fields and their Application to the Sq Field of 1932–33*, Phil. Trans. Roy. Soc. *A256*, 31–98.

RUHE, A., and WEDIN, P. A. (1980), *Algorithms for Separable Nonlinear Least-squares Problems*, SIAM Rev. *22*, 318–337.

(Received January 31, 1990, accepted February 15, 1990)

PAGEOPH, Vol. 134, No. 4 (1990)

0033–4553/90/040533–08$1.50 + 0.20/0

The Effect of the Oceans and Sedimentary Cover on Global Magnetovariational Field Distribution

A. V. Kuvshinov,[1] O. V. Pankratov[1] and B. Sh. Singer[1]

Abstract — Electromagnetic fields excited by long-period geomagnetic variations were calculated for spherical earth models with the realistic inhomogeneous surface layer. Calculations were also carried out for the model with the double inhomogeneous layer. The modelling results did not display the same level of distortions which had been observed experimentally. The revealed contradiction may be explained by a possible existence of significant inhomogeneities in the earth mantle.

Key words: Electromagnetic induction, nonuniform earth, numerical modelling.

1. Introduction

The interpretation of global magnetovariational observations is usually based on one-dimensional inversion. The assumption of the conductivity distribution being laterally uniform appears valid for the earth structure at a depth greater than 400 km. Regional soundings show that at a shallower depth the structure is not laterally uniform. The problem of the period one should use to ignore these inhomogeneities is a central issue in deep geomagnetic sounding. Certainly the answer depends upon the structure of the earth. Thus one should start from the construction of a model, being as realistic as possible.

2. The Model Construction

One of the most advanced models of the earth which presently can be reliably constructed consists of a nonuniform surface layer and some underlying layered medium. We now have positive information about the surface layer, which includes oceans and the sedimentary cover of continents, the conductance S of this layer is known. A number of global soundings provides us with information regarding the

---

[1] Institute of Terrestrial Magnetism, Ionosphere and Radiowaves Propagation of the USSR Academy of Sciences (IZMIRAN), 142092, Troitsk, Moscow Region, USSR.

well-conducting part of the underlying structure. In accordance with all global models it occupies depths greater than several hundred kilometers. The upper part of the underlying structure is highly resistive, it may be properly characterized by a transverse resistance

$$T = \int_{r_0 - h}^{r_0} \sigma^{-1}(r)\, dr, \tag{1}$$

where r_0 is the earth radius, h is the resistive layer thickness, r is the distance from the earth center.

The transverse resistance value is an extremely important parameter of the underlying structure as it determines the value of leakage currents. It cannot be immediately obtained from the magnetic field observations. This follows from the fact that the magnetic field in nonconductive atmosphere belongs to the inductive mode being weakly sensitive to parameters of high-ohmic layers.

Nevertheless it was shown in (FAINBERG *et al.*, 1990b) that the resistive layer manifested itself in the magnetic field indirectly. It was evaluated in their paper that the Sq-variation data were consistent with the averaged value of the transverse resistance $T \approx 3 \cdot 10^9$ Ohm · m². Thus we proceed with the spherical model with an inhomogeneous surface layer used to represent the oceans and the continental sediments. Numerical calculations are carried out with the surface conductance distribution published in (FAINBERG and SIDOROV, 1978). The resistive layer has conductivity σ_T and the thickness h_T. The deeper medium has conductivity distribution

$$\sigma(r) = \sigma_0 \cdot \left(\frac{r_0 - r}{h_0}\right)^{\gamma}, \tag{2}$$

where $h_0 = 4 \cdot 10^5$ m, $\sigma_0 = 1.5 \cdot 10^{-2}$ S/m and $\gamma = 4.735$ in accordance with (FAINBERG, 1983). Values of h_T and σ_T are chosen to satisfy conditions $T = h_T/\sigma_T$ and $\sigma_T = \sigma_0(h_T/h_0)^{\gamma}$.

To get the magnetic field induced in the model by an external field we first calculate currents $\mathbf{j}^s$ flowing in the surface nonuniform layer. These currents satisfy the integral equation (FAINBERG *et al.*, 1989a)

$$\mathbf{j}^s(r) = \mathbf{j}_0^s(r) - \int_s \hat{G}_0(r, r')\, \frac{R^*}{R_0}\, \mathbf{j}^s(r)\, ds'. \tag{3}$$

Here R_0 is some constant value, we call R_0^{-1} the reference surface conductance. The laterally uniform model differing from our original one by substitution of the nonuniform surface layer with a uniform one, whose conductance is R_0^{-1}, we call the reference model; $\hat{G}_0(r, r')$ is the Green's tensor function of the reference model, r is the pair of surface coordinates $\{\theta, \varphi\}$,

$$R^*(r) = S^{-1}(r) - R_0. \tag{4}$$

The term $\mathbf{j}_0^s$ is the surface current that would be excited in the reference model by the external magnetic field being under consideration.

Equation (3) can be solved in a wide frequency band with the help of an iterative-dissipative method (FAINBERG et al., 1990a).

3. The Numerical Calculations

Calculations have been carried out for the uniform external magnetic field described by the only spherical harmonic $P_1^0(\theta, \varphi)$ at the set of periods 12 h, 1, 3, 10 and 15 days. Models with transverse resistance T equal to 10^8, $3 \cdot 10^9$ and $7 \cdot 10^9$ Ohm $\cdot$ m^2 as well as the model with an intermediate layer of conductance 2000 S, situated at depth 50 km, have been considered. Results for periods of 1 and 15 days and the resistive layer with $T = 3 \cdot 10^9$ Ohm $\cdot$ m^2 are shown on Figures 1–3. The values plotted are the apparent resistivity ρ_a and phase of the impedance $Z_1 = -i\omega\mu_0\lambda_1$ (scaled by factor C given in the figure caption), where

$$\lambda_1 = -\frac{r_0}{2} \cdot \frac{H_r}{H_\theta} \cdot \tan\theta \tag{5}$$

and H_r, and H_θ are the vertical and azimuthal magnetic fields components. It can be seen from Figure 1 that for the 1 day period ρ_a differs from the normal value $\rho_a^n = 49.6$ Ohm $\cdot$ m. It is 31% smaller in Japan, 21% greater on the western shore of North America, in the southern Australia and Indo-China, 97% greater in South Africa, where a significant current concentration takes place. The phase shift (Figure 2), with respect to a normal value $\varphi^n = -76°$, reaches $-39°$ to $+16°$. For the 15 day period the apparent resistivity (Figure 3) is much closer to the normal value $\rho_a^n = 7.4$ Ohm $\cdot$ m. The maximum shift in this case is equal to 9%. It should be noted here that extremely significant distortion takes place beside the equator where H_r is small. Therefore, to make plots more readable we extracted the data inside the strip $\theta = -10°$ to $+10°$ before plotting.

4. Discussion

The calculations show that surface inhomogeneities cause the distortions of the apparent resistivity distribution. The distortions are more significant at smaller periods. The corresponding empiric results were obtained by a number of authors. We can refer for example to the well-known results of ROBERTS (1982). Comparison shows that our results cannot explain more significant distortions observed experimentally. This means that some additional and more prominent inhomogeneities should exist in the earth's crust and mantle. Correspondingly, electromagnetic fields induced in more sophisticated models should also be calculated.

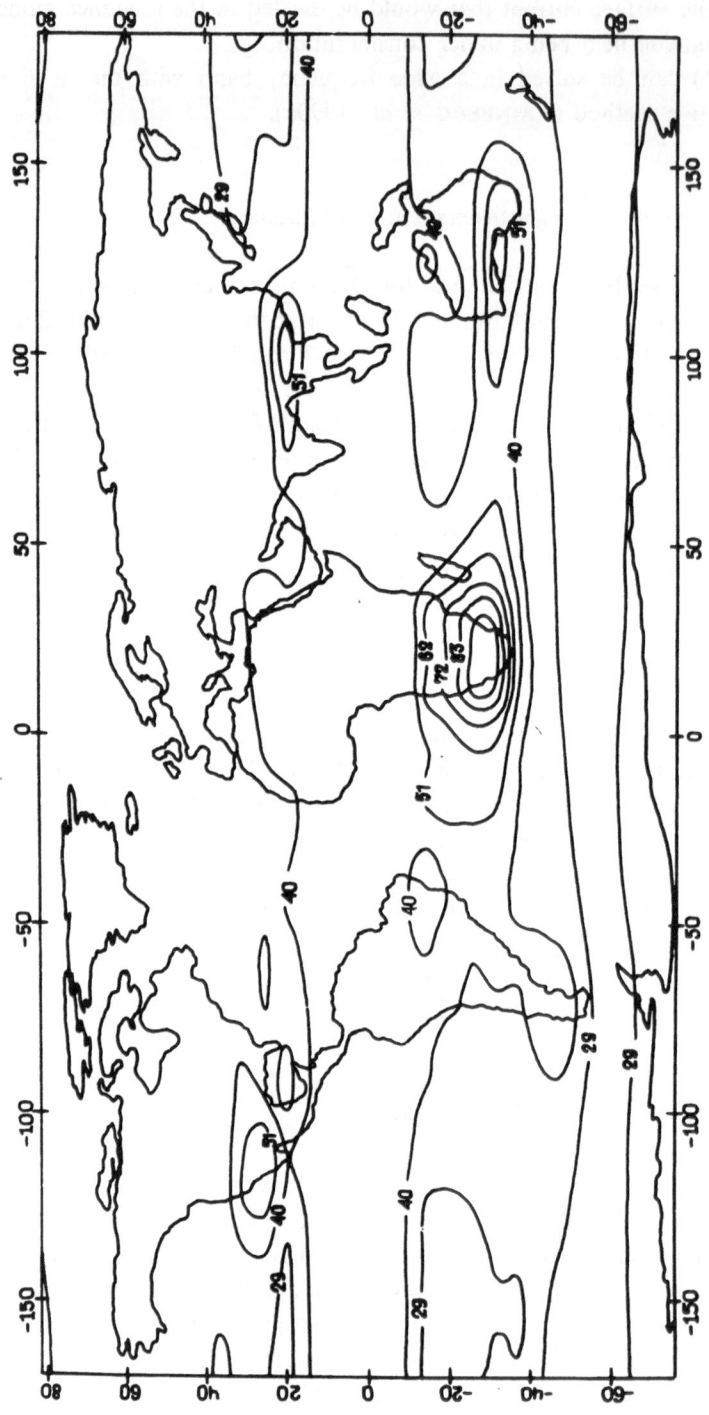

Figure 1

Apparent resistivity for period 1 day (model 1); $C = 1.177$ Ohm · m.

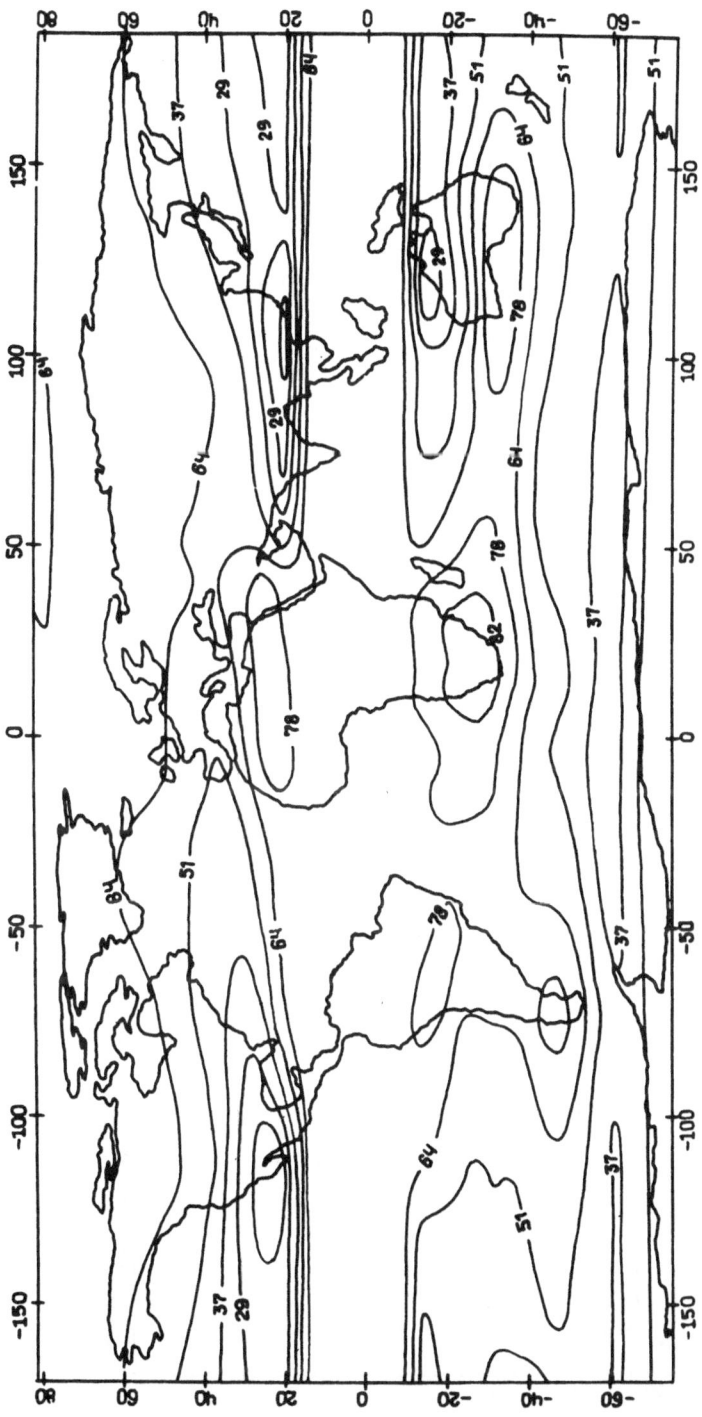

Figure 2

Phase of the impedance for period 1 day (model 1); $C = -1.094°$.

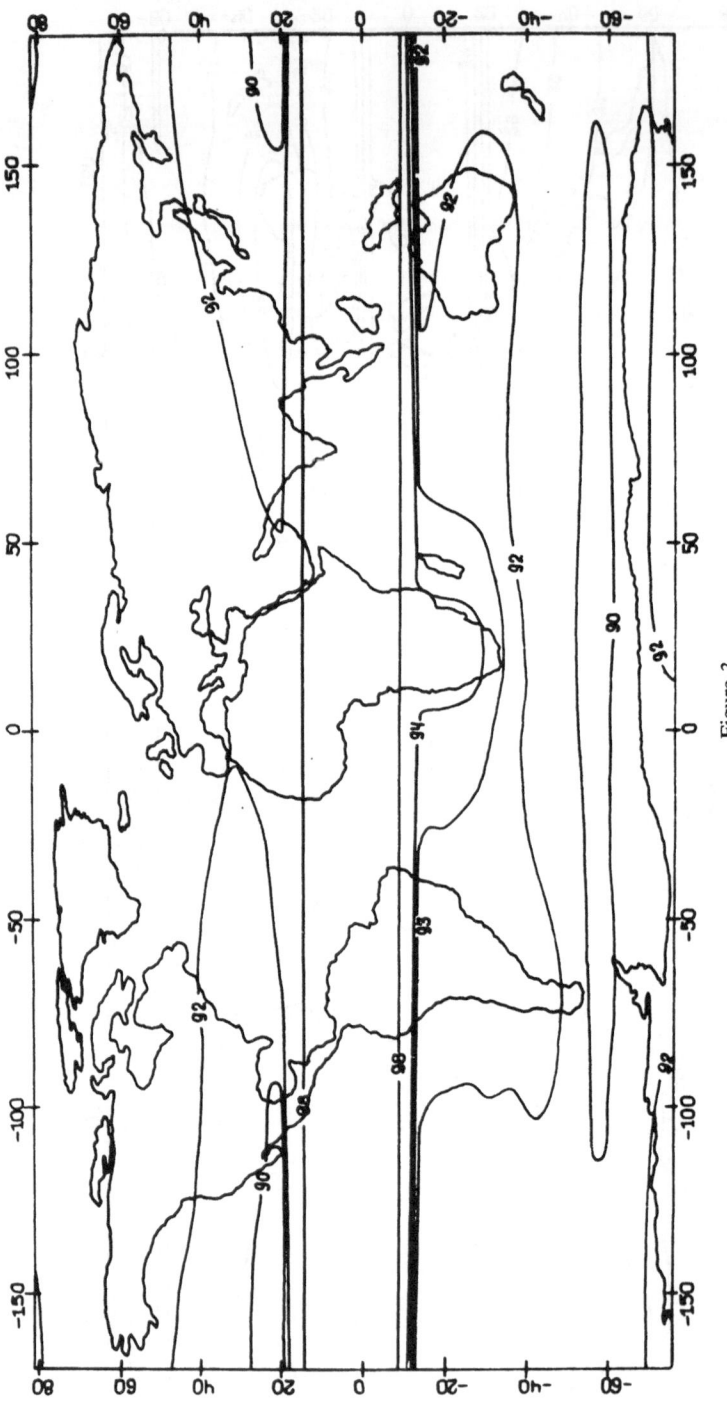

Figure 3

Apparent resistivity for period 15 days (model 1); $C = 0.075$ Ohm · m.

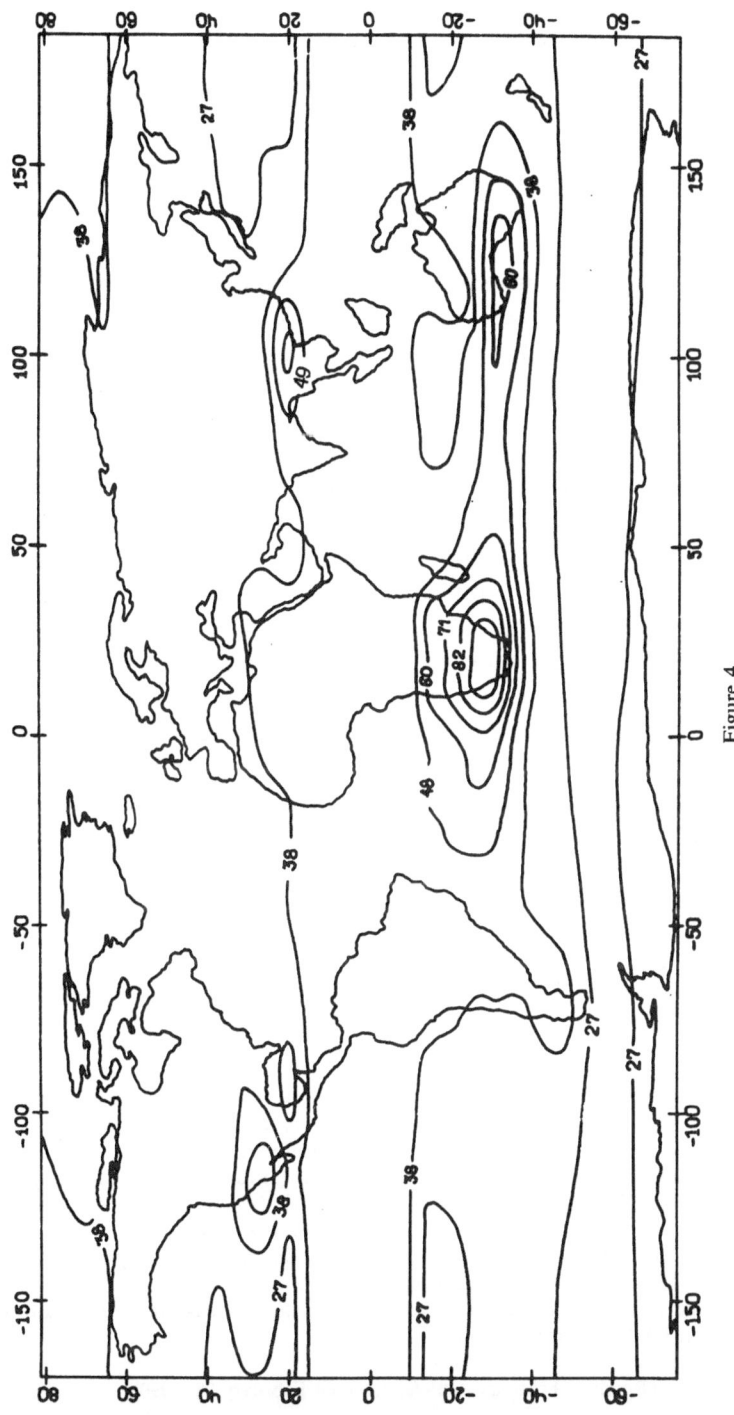

Figure 4

Apparent resistivity for period 1 day (model 2); $C = 1.174\ \mathrm{Ohm \cdot m}$.

Even now a further step can be made. We can consider a model whose resistive layer is also inhomogeneous. In this case equation

$$\mathbf{j}^s(r) + \mathbf{n}j_n(r) = \mathbf{j}_0^s(r) - \int_s \hat{G}_0(r, r')\left\{\frac{R^*}{R_0} \mathbf{j}^s(r') + \frac{T^*}{R_0} \mathbf{n}j_n(r')\right\} ds' \qquad (6)$$

instead of eqn. (3) should be solved (SINGER and FAINBERG, 1985). Here

$$T^*(r) = T(r) - T_0, \qquad (7)$$

T_0 is the base level of the transverse resistance, $\mathbf{n}$ is the vertical unit vector and $\mathbf{n}j_n$ is the density of vertical leakage currents flowing through the resistive layer. The equation can also be solved using the iterative-dissipative method.

Figure 4 which displays ρ_a has been obtained for the 1 day period in this model. The resistive layer has resistance 10^8 Ohm $\cdot$ m^2 under the oceans and 10^{10} Ohm $\cdot$ m^2 under the continents. Calculations reveal significant distortions, however they cannot explain distortions which have been experimentally observed, so we can summarize that even this modified model is not a satisfactory one.

The results which have been obtained should be considered as preliminary ones. Some uncertainty still exists due to the fact that comparably rough numerical mesh with the cell size $5^0 \times 5^0$ has been used. To make the final conclusions one needs to incorporate additional information concerning the behaviour of surface inhomogeneities along a smaller space scale in the vicinity of magnetic observatories. The refining calculation can be carried out for the plane models and also with the use of thin sheet models. The global scale field in this case may be used as a normal field. Nevertheless the current calculations draw us to the conclusions that observations cannot be explained without taking deeper inhomogeneities into account.

REFERENCES

FAINBERG, E. B., and SIDOROV, V. A. (Eds.), *The Total Longitudinal Conductance of Sedimentary Cover and Water Shell of the Earth* (Nauka, Moscow 1978) 15 pp. (in Russian).
FAINBERG, E. B., *Global geomagnetic sounding*, In *Mathematical Modeling of Electromagnetic Fields* (IZMIRAN, Moscow 1983) pp. 79–121 (in Russian).
FAINBERG, E. B., KUVSHINOV, A. V., and SINGER, B. Sh. (1990a), *Electromagnetic Induction in a Spherical Earth with Nonuniform Oceans and Continents in Electric Contact with the Underlying Medium: I—Theory, Methods and Example*, Geophys. J. (in press).
FAINBERG, E. B., KUVSHINOV, A. V., and SINGER, B. Sh. (1990b) *Electromagnetic Induction in a Spherical Earth with Nonuniform Oceans and Continents in Electric Contact with the Underlying Medium: II—Bimodal Global Geomagnetic Sounding of the Lithosphere*, Geophys. J. (in press).
ROBERTS, B. Sc. (1982), *The Electromagnetic Response of the Earth and Upper Mantle Electrical Conductivity*, Ph.D. thesis, University of Lancaster, 125 pp.
SINGER, B. Sh., and FAINBERG, E. B., *Electromagnetic Induction in Non-uniform Thin Layers* (IZMIRAN, Moscow 1985) 234 pp. (in Russian).

(Received February 3, 1990, accepted February 28, 1990)

PAGEOPH, Vol. 134, No. 4 (1990)

0033–4553/90/040541–17$1.50 + 0.20/0

The Magnetospheric Disturbance Ring Current as a Source for Probing the Deep Earth Electrical Conductivity

WALLACE H. CAMPBELL[1]

Abstract—Two current rings have been observed in the equatorial plane of the earth at times of high geomagnetic activity. An eastward current exists between about 2 and 3.5 earth radii (Re) distant, and a larger, more variable companion current exists between about 4 and 9 Re. These current regions are loaded during geomagnetic substorms. They decay, almost exponentially, after the cessation of the particle influx that attends the solar wind disturbance. This review focuses upon characteristics needed for intelligent use of the ring current as a source for induction probing of the earth's mantle. Considerable difficulties are found with the assumption that *Dst* is a ring-current index.

Key words: Magnetosphere, ring current, *Dst* index, earth conductivity.

1. Introduction

Many years have passed since the first studies using electromagnetic fields from natural sources probed the earth's internal electrical conductivity (SCHUSTER, 1890; CHAPMAN, 1919; CHAPMAN and WHITEHEAD, 1922; CHAPMAN and PRICE, 1930; LAHIRI and PRICE, 1939; CHAPMAN and BARTELS, 1940). Physical laws define the depth to which such conductivity profiles can be obtained; the longer period field variations go deeper into the earth; penetration decreases with increasing conductivity of the medium. For the earth's interior structure that shows a generally increasing conductivity with depth below the crust, studies indicate that the *Sq* diurnal variation of field only penetrates to about 500 or 600 km depth (CAMPBELL and SCHIFFMACHER, 1988). Other natural signal sources with field change periods longer than one day are needed to reach deeper. The magnetospheric ring current, having spectral components that extend to several days or more can be the source for electrically exploring the earth's lower mantle. The purpose of this review is to consider the characteristics of the ring current from the viewpoint of the geophysicists' need for a well-behaved, electromagnetic signal source of a very long period.

Our principal knowledge of the earth's deep interior composition arises from four roots: (1) seismic observations providing earth density profiles; (2) surface

---

[1] U.S. Geological Survey, Mail Stop 968, Box 25046, Denver, CO 80225, U.S.A.

geological evidence of materials from the earth's deep interior; (3) extrapolation of the cosmic origin models of the earth using meteoric samples found on the earth; and (4) laboratory research using a high-temperature, high-pressure environment on possible mantle materials. The most definitive seismic records seem to indicate density transitions near 400, 650, and 1050 km into the earth's mantle (DZIEWON-SKI and ANDERSON, 1981). An active research endeavor has been the search for an understanding of the composition and phase changes that cause these transitions. A determination of the electrical conductivity of the region, obtained independent of the seismic evidence, can contribute to the accurate definition of the earth's mantle.

The earth's main field arises from motions of the fluid outer core of the earth. These motions are responsible for the unique dipole and multipole structure, westward drift, secular change, and paleomagnetic reversals. The electromagnetic coupling between the outer core and the earth's mantle needs to be understood for relating the core processes to the geomagnetic field observations at the earth's surface. An accurate conductivity profile of the mantle is necessary for proper evaluation of this coupling.

The earth's temperature and pressure increase rapidly with depth. The electrical conductivity of an earth-embedded material, of a given composition and phase, rises about exponentially with the negative reciprocal of the temperature (TOZER, 1970). Phase changes, occurring as the pressure increases with depth, cause stepwise density and conductivity transitions at unique depths. Attempts to determine the electrical conductivity within the deep earth have provided diverse results over the past years (RIKITAKE, 1950, 1966; MCDONALD, 1957; CANTWELL, 1960; YUKU-TAKE, 1965; BANKS, 1969, 1972; PRICE 1973; FAINBERG and ROTANOVA, 1974; DMITRIEV et al., 1977; ISIKARA, 1980; FILLOUX, 1980; KOVTUN and PROKHOVA, 1980; ROKITYANSKY, 1982). At the present time, the general expectation of the earth conductivity change with depth below the crust seems to be a gradually rising value from about 0.01 to 0.1 S/m. Then a rapid rise may occur near 400 or 500 km, giving an order of magnitude increase in conductivity over a 50–100 km transition zone. Next, some response to the 650 km density step is anticipated. Past that transition, the conductivity is expected to rise gradually toward a value of about 1.0–3.0 S/m until the density transition near 1050 km is reached. However, this expectation is not strongly supported with analytical determinations. A careful use of the magnetospheric ring current as a probe source may be the key to this effort; but without detailed knowledge and precise utilization of this source behavior, the accuracy of any conductivity computations can be compromised.

2. Geomagnetic Storms and Substorms

There was a time, not too long ago, when many earth scientists believed that all the large field variations indicated on a magnetic recording on non-quiet days were due to a simple current of particles encircling the earth like a ring of Saturn. Perhaps, because such a source geometry simplifies earth conductivity computations, some solid earth geophysicists still tenaciously hold to this outdated viewpoint. In just the past few decades, space physicists have shown that the disturbed geomagnetic field variations are a superposition of many upper atmospheric and magnetospheric processes. Below I will briefly summarize some of the present knowledge of the ring current characteristics that should be of interest for induction studies.

Dynamic processes on the Sun deliver a plasma of charged particles (principally protons and electrons) and associated fields to the earth's environment, causing geomagnetic disturbances at the earth's surface which have been named "geomagnetic storms." At midlatitudes on the earth, about one storm a year is larger than 250 gammas (1 gamma $= 1$ nT) and about 10 per year are over 50 gammas. The number and intensity of disturbances vary with the 11-year solar activity cycle with about a one-year lag. At middle and low latitudes on the earth, many storms display a similar general appearance (Figure 1), although some of the "average" features occasionally may be absent. First, there is a *Sudden Commencement* (SC) that occurs almost simultaneously (within minutes) everywhere on earth and is followed by a general increase in the magnetic northward field (*Initial Phase*) that continues for as long as several hours. Next, follows the *Main Phase* of the storm

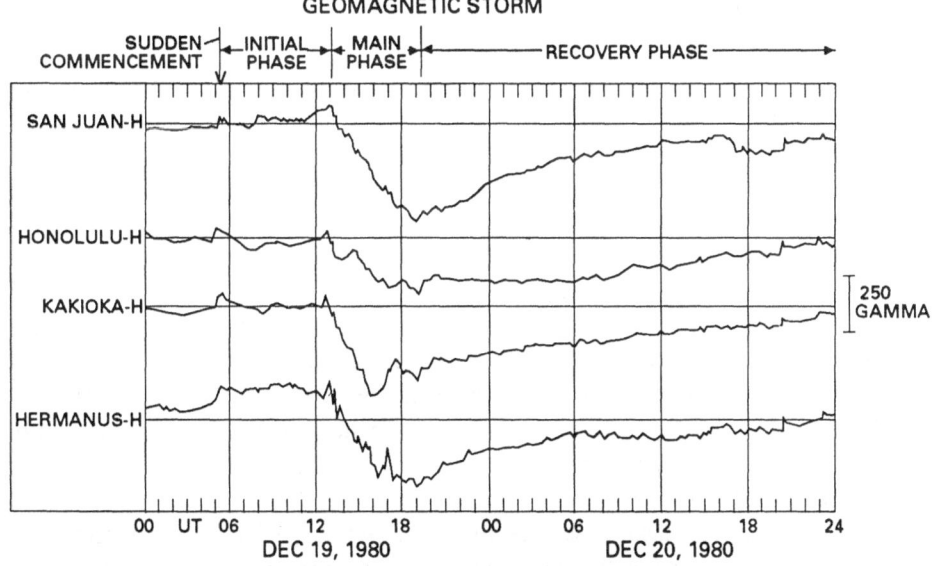

Figure 1

Characteristics of a typical geomagnetic storm shown in the *H* component of field variation at the four stations indicated during the *UT* hours of 19 and 20 December 1980. These stations are the ones used for derivation of the present *Dst* index.

in which this principal component of the field decreases and shows major fluctuations for a time lasting a little longer and with larger amplitude than the initial phase. Finally, in a *Recovery Phase*, the storm spends its longest time gradually returning to its undisturbed level over as much as several days. In general terms, the solar wind drives a convection of the magnetosphere which dissipates in the geomagnetic storm-related processes. The storm energy is divided between the input to the auroral ionosphere, the creation of the ring current, and the magnetospheric down-tail processes.

The storm SC's are related to the arrival of the solar wind hydromagnetic shocks at the geomagnetic field interface (stand-off point); the SC size is roughly proportional to the square root of the solar wind dynamic pressure. The Main Phase depends upon a sustained southward-directed interplanetary magnetic field ($-Bz$ IMF) of the solar wind at the magnetospheric boundary. The SC size is independent of the Main Phase and not all SC's are followed by a storm Main Phase. During the storm Main Phase, there are major increases in the plasma of trapped ions and electrons between $L = 2$ and $L = 9$. (L is, roughly, the equatorial distance of the dipole field that is equivalent to the actual main field of the earth, in earth-radii units, Re). With the injection of plasma into the magnetosphere, a charge separation arises across the dawn and dusk sides of the magnetosphere that drives field-aligned current systems which transfer energy between the magnetosphere and the ionosphere. The equatorial drift of the injected particles about the earth forms a ring current, causing a magnetic field decrease at the earth's surface. The Recovery Phase of the storm is related to a decay of the ring current system attending the "switching-off" of the $-Bz$ IMF.

The group of related geophysical processes that recur during the storm's initial and main phases are called *Polar Substorms*. As the high-speed (500–870 km/sec) matter from the sun arrives at the earth's outer field lines (typically near 12 earth radii, Re) and then compresses the earth's magnetosphere to as much as 5 Re, it carries along a specific IMF direction. The compression causes an increase in the northward directed field at the earth's surface. If the IMF field, arriving with the sun-ejected plasma, is directed southward (allowing a field line connection between the solar wind and the northward directed magnetospheric field) a major magnetic substorm can be triggered in which the arriving particles enter and modify the shape and composition of the magnetosphere. If the IMF is strong and northward ($+Bz$), small, complicated, field-aligned currents (called the NBZ system) flow into and away from the sunlit polar cap ionosphere. Currents in the polar region are quite sensitive to the IMF direction toward (T) or away (A) from the sun. Because this alignment occurs in distinct sectors about the sun, a special "Sector Effect" has been recognized in geomagnetic measurements. The spiral angle of the solar wind arriving at the magnetosphere causes the $+By$ IMF to be related to the A sector and $-By$ IMF to the T sector geomagnetic field effects. The polarity of the By IMF has been identified with the global north-south and dawn-dusk asymmetries of the magnetospheric current patterns. The solar wind speed, together with the magnitude and direction of the IMF, control the growth and decay of the substorm events.

During typical substorms, there is an increase in the magnetic flux in the lobes of the tail and in the east-west cross-tail current, and the tail plasma sheet thins near the earth. Currents flowing from the tail region feed the growth of a ring current about the earth. Energized particles, usually electrons, then precipitate into the nightside ionosphere causing aurora and a massive flow of field-aligned (Birkeland) currents into and away from the auroral region. Other particles from the arriving solar wind enter the cusp of the magnetosphere, a region defined by the

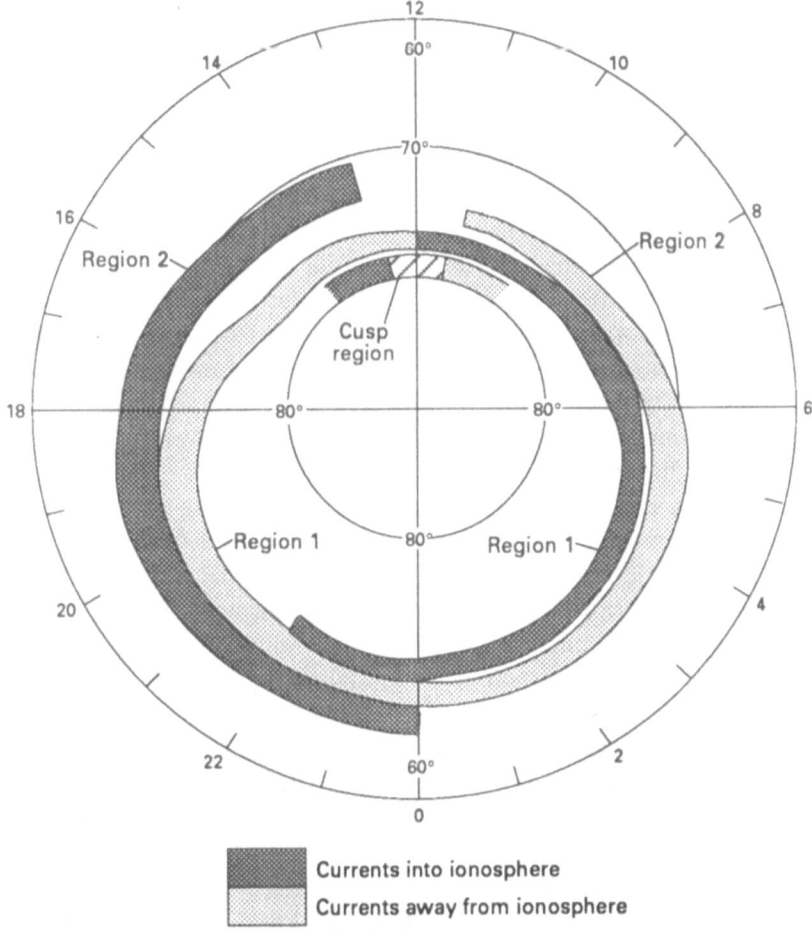

Figure 2

Region 1 and region 2 field-aligned (Birkeland) currents in the polar region. Circles of invariant latitude (see text) and local time are indicated. Figure redrawn from IIJIMA and POTEMRA (1976).

division between the magnetospheric field lines on the sunward and the downwind (anti-sunward) side of the earth.

Figure 2 illustrates the field-aligned (Birkeland) current system during a typical auroral substorm (IIJIMA and POTEMRA, 1976). The inner, region 1, current systems have been associated with particles of the magnetospheric boundary layer. In this Northern Hemisphere view, region 1 current flows into the auroral zone on the morning side and away from the auroral zone on the afternoon side. The outer, region 2, currents flow in directions opposite the region 1 system and seem to be connected with the ring current. Small, field-aligned currents associated with the cusp region are shown in Figure 2 poleward, near the noon meridian. The locus of all these currents projected onto the ionosphere define the "auroral zone" (sometimes called the "auroral oval").

During a substorm, in the auroral nightside ionosphere, a westward electrojet current, driven by the Birkeland currents, flows in the region of enhanced conductivity created with the aurora. This strong westward electrojet current depresses the northward geomagnetic field at those latitudes with violent variations in synchronization with the overhead aurora. Very short-period (seconds to minutes) field pulsations, and associated auroral luminosity fluctuations are measured at this time. The auroral region of maximum field disturbance is generally restricted to less than 5° in latitude and 100° in longitude on the nightside of the earth. The ratio of high to low frequency components of the geomagnetic disturbance decreases rapidly with distance from this region and with level of activity. Approximately an hour after the auroras and electrojet currents of a substorm have reached their peak development, there is typically a "poleward leap" of these phenomena to very high latitudes, in association with a magnetotail plasma sheet thickening and downwind retreat of the tail current sheet; the substorm ends.

The conducting ionosphere allows a closure of the strong substorm westward electrojets to be communicated to other longitude and lower latitude locations. At typical mid-latitude locations on the earth's surface, the varying fields of the complex magnetospheric and ionospheric substorm currents may be mixed, as if randomly taking turns dominating the observations with contributions that depend upon each individual source magnitude and distance. Because the principal current-carrying E and lower F regions of the ionosphere are just about 100–200 km altitude, fields from these nearby source regions are a major part of most earth surface observatory records.

From the very early stages of the substorm development, electrons and ions are injected into a ring-like region about 2.5–8.5 Re distant in the geomagnetic equatorial plane of the earth. The ring current is fed by field-aligned currents connecting to the tail region of the magnetosphere and to the ionosphere. The field and particle interactions actually generate two oppositely directed ring current regions. The outer, westward flowing ring dominates the disturbance contributions to the field at the earth with a southward, axially-directed world-wide component.

This ring current field reaches its maximum value during the superposed substorm contributions to the Main Phase of the geomagnetic storm. The slow decay seen in the storm Recovery Phase represents the decay of the ring current and its associated Birkeland currents and lasts from several hours to several days.

3. Ring Current Characteristics

Although measurements indicate that the magnetospheric ring current is an ever-present feature of the magnetosphere, even in storm times the relative magnitude of the perturbations on the earth's strong geomagnetic dipole field by the ring is small (Figure 3b). The relative size and position of the eastward and westward current (positive ion flow direction) systems of the ring are shown in Figure 3a. Figure 4 illustrates four satellite-pass observations (superposed) of the ring current for two storm periods. Note that the consistent eastward currents are between L-values of about 2 and 3.5. The dominant westward current flow occurs between L-values of about 3.5 and 8 (the limit of the satellite sampling altitude). Both spatial and temporal fluctuations seem to persist in the westward system during the height of the ring current activity; the eastward current, also enhanced during a storm, is considerably less variable.

A significant asymmetry of the ring current occurs during the height of the storm with intensities on the nightside about 2–3 times those on the dayside. The ring is most symmetric during the storm recovery phase attending the demise of the substorm particle injection. With the changing season, as the geomagnetic tail moves to the anti-solar position, the ring current deforms slightly as an accommodation to its location in the earth's deformed dipole field lines. Region 2 Birkeland currents have been observed connecting the westward ring current to the auroral ionosphere (Iijima et al., 1990). These currents seem to flow toward the earth between about 16 and 24 hr local time and away from the earth between 2 and 7 hr local time. The typical ionospheric position of the substorm field-aligned currents is illustrated in Figure 2 at 70° invariant latitude, Λ (Λ an angle from the relationship $\cos^2 \Lambda = 1/L$; i.e., at $L = 8.5$).

Figure 5 illustrates the composition, number density, and energy density in the ring current obtained from satellite measurements. Low energy protons (H^+ at about 15–250 keV) make up about 70 per cent of the composition throughout the bulk of the ring current, but oxygen and helium are in significant abundance also. Some researchers believe that the oxygen of low charge state is of ionospheric origin. The largest westward current can be several Re distant from the location of the maximum in population density. However, the magnitude of these field variations at the earth's surface is found to be proportional to the total kinetic energy of the particle population. It requires about 4 K Joules of particle energy to create 1 gamma of field at the earth's surface.

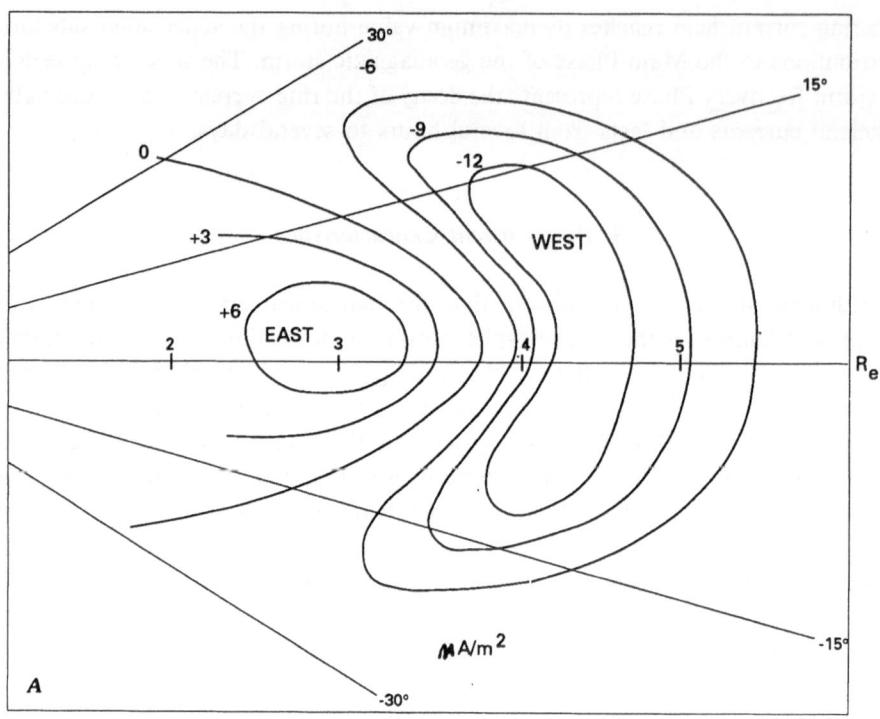

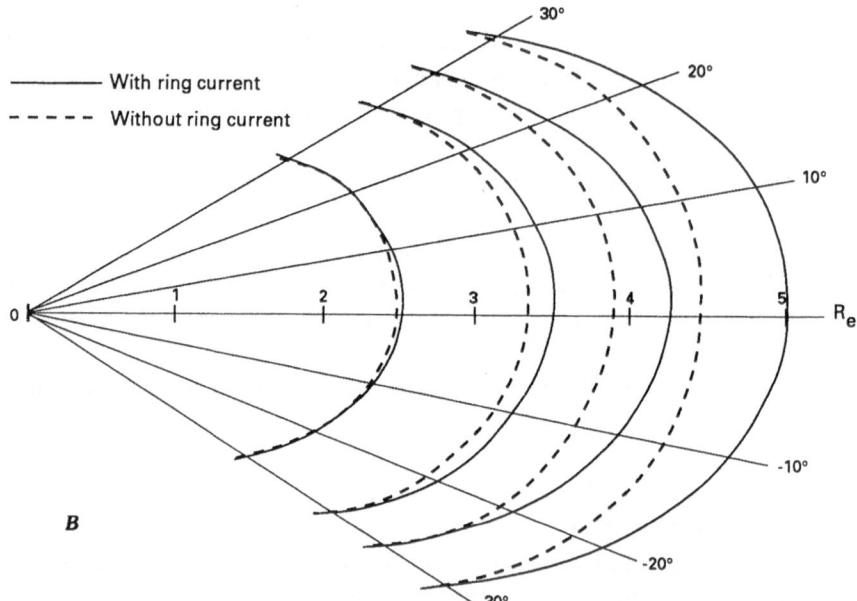

Figure 3
(a) North-south cross section of ring current profile, (b) dipole field line distortions caused by the ring current. Figure redrawn from HOFFMAN and BRACKEN (1967).

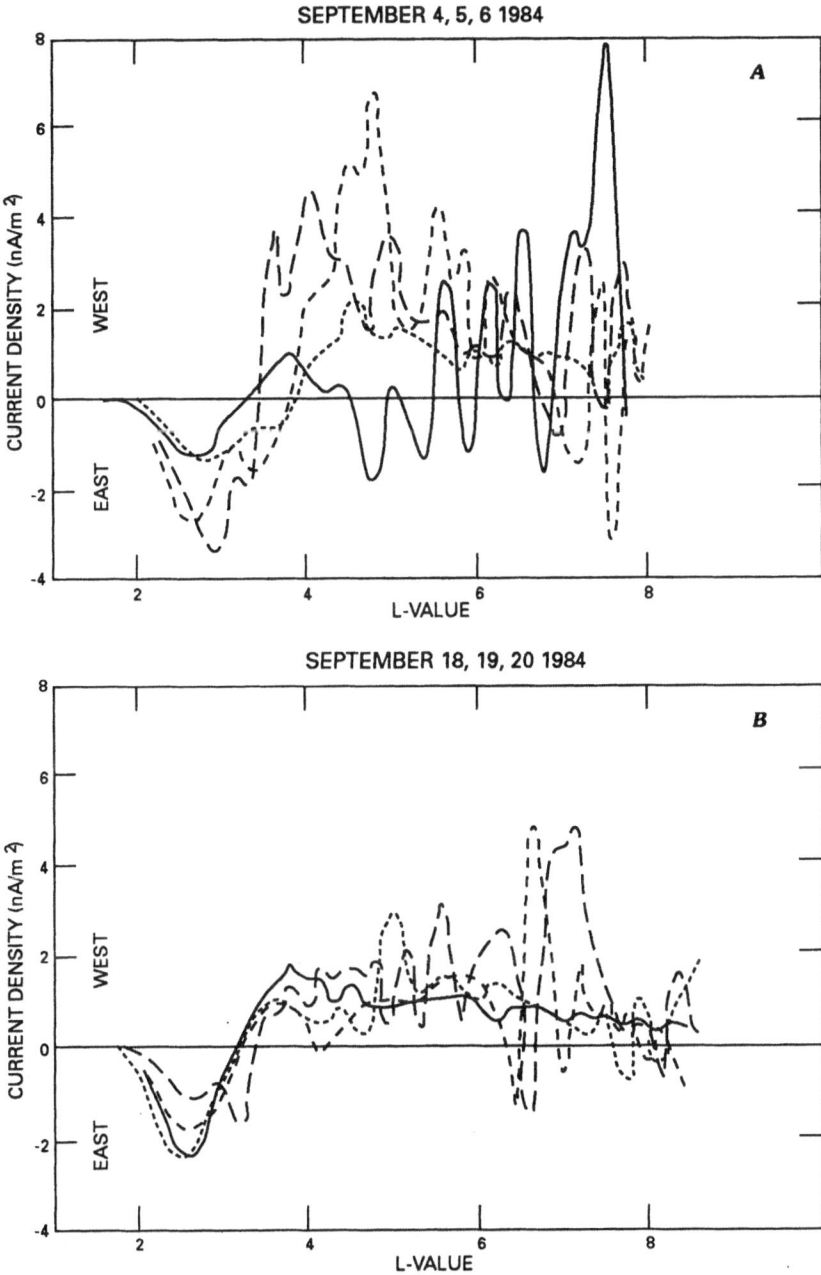

Figure 4

Radial profiles of ring current densities during two storms. September 4–7 and September 18–20, 1984. Four passes of the AMPTE/CCE satellite are overplotted in each example. Current density units given to left; radical distance in *L*-shell units (see text) given at bottom. Figure redrawn from LIU *et al.* (1987).

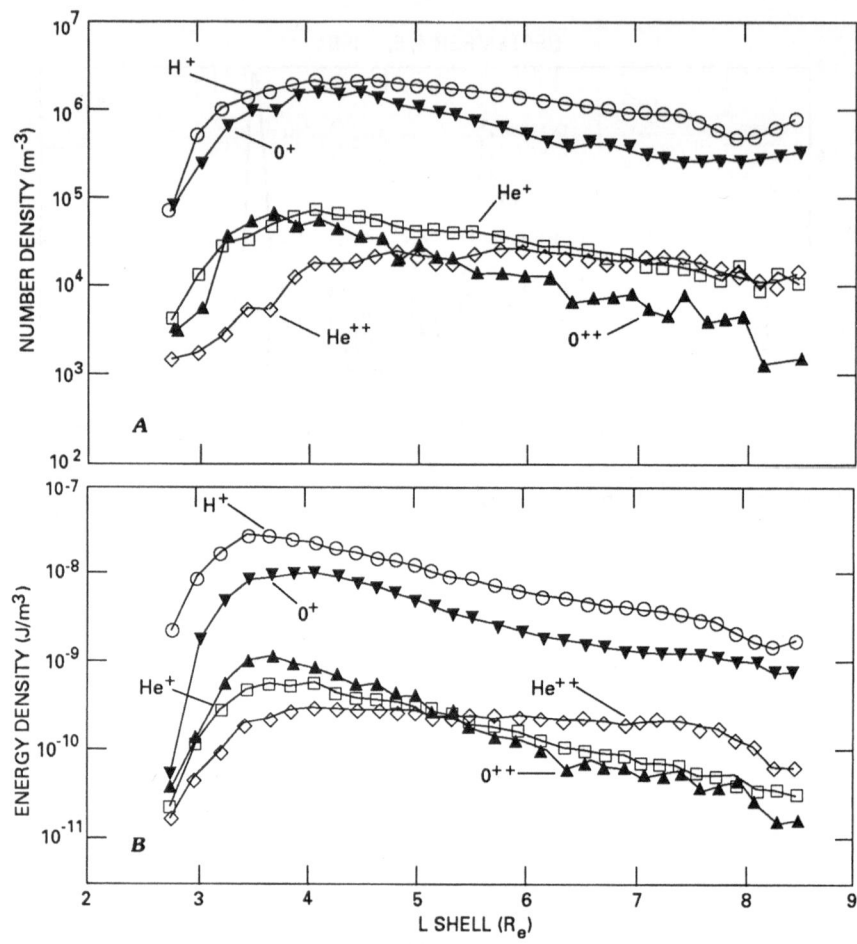

Figure 5

Radial profiles of (a) number density and (b) energy density during the AMPTE/CCE satellite pass of 5 September 1984. Number density and energy density units indicated at left; radial distance in L-shell units (see text) given at bottom. Values are given separately for protons (H^+), Oxygen (O^+ and O^{++}), and Helium (He^+ and He^{++}). Note difference in L-shell of maximum for (a) and (b). Figure redrawn from ROELOF and WILLIAMS (1988).

An axially symmetric ring of westerly current surrounding the earth adds a southward, axially directed field to the earth's surface measurements. This contribution decreases the magnetic northward directed vector measurement by an amount $H \cos(90 - \theta)$, where θ is the geomagnetic colatitude, and changes the Z component by an amount $Z \sin(90-\theta)$. The ring current field, together with the substorm fields from the tail current, Birkeland currents, auroral electrojet, and ionospheric closing currents all contribute to the H-component depression seen during the Main Phase period of the magnetic storm (Figure 1). When the substorm injection of particles drops below the decay rate (governed by the charge-exchange collision of ring current protons with cold hydrogen atoms of the magnetosphere) the Recovery Phase of the magnetic storm starts. The observed field decay rate is roughly proportional to the strength of the ring current in the absence of sources.

With moderate to large geomagnetic storms, when the magnetic activity index, Kp, is 5 or greater, mid-latitude stable auroral red (SAR) arcs of 6300 Å oxygen emissions have been detected at an altitude coincident with an $F2$ ionospheric region electron density depression at 300–600 km altitude (REES and ROBLE, 1975). Inside an arc region, electron temperatures are noticeably enhanced with respect to the surroundings. These upper atmosphere luminosity emissions, almost symmetrically located in the Northern and Southern Hemispheres on the dark side of the earth, are distinctly separated by several degrees from the higher latitude, simultaneously occurring auroras. SAR arcs typically seem to align, within 1°, along magnetic invariant latitude, Λ, contours, usually near $\Lambda = 45°$ to 55° (Figure 6). Such locations correspond to the magnetospheric locus of the eastward ring current (cf. Figure 4). The fact that no magnetic perturbation occurs at locations beneath the SAR arc (unlike auroras) is probably related to the absence of E-region ionospheric conductivity on these nightside regions of the earth; at similar invariant latitudes on the dayside there should be surface fields from ionospheric currents driven from the auroral region substorm system. The relative purity of the red line 6300 Å emission implies low energy source bombardment of the atmosphere that does not penetrate to E-region altitudes. There is a higher oxygen to proton density ratio

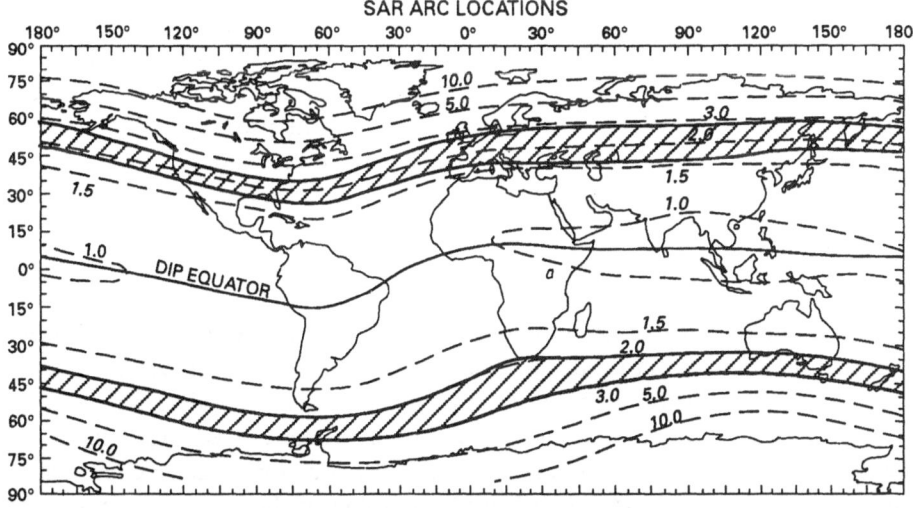

Figure 6

OGO satellite 6300 Å airglow observations for the 28–29 September 1967 SAR arc positions superposed upon L-shell (see text) contour lines. Arc position attributed to Reed and Blamont shown in REES and ROBLE (1975) redrawn on L-shell map.

inside the arc than outside. It seems clear from the timing and location that the SAR arcs are an ionspheric manifestation of the inner, magnetospheric eastward ring current.

4. The Dst Ring Current Index

The *Dst* is a geomagnetic index (MAYAUD, 1980; RANGARAJAN, 1989) representing the global variations of the northward field components that have been normalized to the equator and averaged, in Univeral Time, over all station longitudes (Figure 7). The index has been fixed into a "classical" form that uses hourly measurements of *H* from just a few low-latitude stations, the removal of quite-time daily variations, adjustment for main field changes, and evaluation at all *UT* hours. *Dst* has come to be known as a "Ring Current" index because the averaging and normalization emphasizes those features of a storm that change simultaneously, parallel to the earth's geomagnetic axis as if from a ring of current circling the earth.

The method of deriving *Dst* (*D* for "disturbance," and *st* for "storm-time") was originated by MOOS in 1910 for systematic analysis of geomagnetic storms. The procedure was later elaborated by CHAPMAN (1927) and became essentially a standardized hourly index during the sunspot maximum of the International Geophysical Year, 1958 (SUGIURA, 1964). Since the IGY period, the index has been widely used in studies of the upper atmospheric and space environment. Some induction specialists have suggested that because *Dst* is a valid representation of a symmetrical ring current in the magnetosphere it should find immediate application as a source for deep-earth conductivity studies. However, the index was fixed into "classical" form over 30 years ago before satellite exploration of the magnetosphere exploded the simplistic views of the geomagnetic storm processes. Below I will review some of the problems with the *Dst* as a ring current index.

Unfortunately, the station distribution used for the Dst index is severely limited. Presently, only Honolulu (Hawaii), San Juan (Puerto Rico), Hermanus (S. Africa), and Kakioka (Japan) are employed (cf., Figure 1). The poor longitude distribution means that ring east/west asymmetry features are severely limiting the index. The lack of equivalent contributing stations in the Northern and Southern Hemisphere means that the ring-current north/south asymmetry and seasonal effects are hidden. If the number and distribution of stations are increased, consideration need be given to three location restrictions: (1) exclusion of the dip equator regions as defined by the width and variation in latitude of the equatorial electrojet current in quite and active times (RASTOGI, 1990); (2) exclusion of the auroral regions using the invariant latitude of the ring current as defined by the SAR arc locus; and (3) avoidance of the seasonal and activity-level variation of the *Sq* current focus locations (described below).

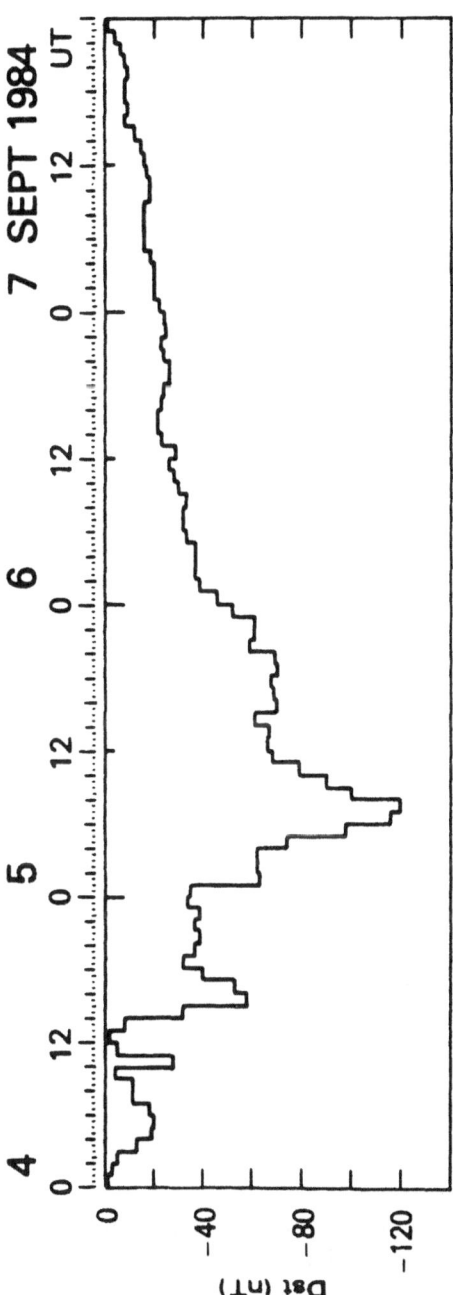

Figure 7

The *Dst* values for the period 4–7 September 1984. Amplitude values are in gammas; Universal Time on the horizontal scale. The index covers the period of one AMPTE satellite measurement of ring current shown in Figures 4 and 5 above. Figure redrawn from WILLIAMS and SUGIURA (1985).

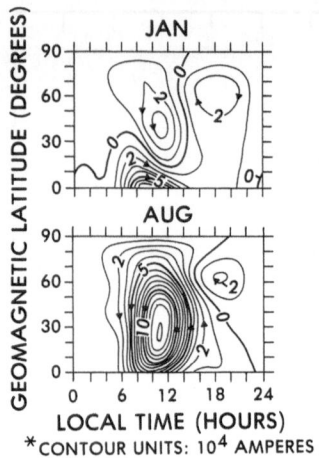

Figure 8

Equivalent ionospheric source currents for Sq daily variations of field in Europe for January and August, 1965. Each pattern, in local-time versus geomagnetic-latitude coordinates, shows 10^4A steps between contours, with arrows for the required flow direction. A zero level at local midnight is assumed. Note the difference in focus location for the two months. A station located between the two foci would show an H-component increase or decrease in the mid-day hours depending upon the season. Other shifts in focus location occur with changes in geomagnetic activity level. Figure redrawn from CAMPBELL (1990).

The daily variation of the field, Sq, on geomagnetically quiet days is removed from the field measurements that are used in the formation of Dst. The purpose is to eliminate the fields of ionospheric dynamo currents from the index. Unfortunately, four problems arise: (1) both the thermospheric wind systems and the E-region ionization are modified during substorm periods so that the dynamo currents in active times are different from those expected by an extrapolation of quiet-time field data before or after the substorm; (2) small changes in the Sq current focus position (TARPLEY, 1973) cause major changes in the phase and the amplitude of the vector field components at observatories within the focus latitude range near midday (Figure 8); (3) sector (By IMF) effects that have been detected in Sq (MATSUSHITA et al., 1973) are not presently accommodated; (4) there are small, semidiurnal lunar tidal dynamo current effects in the present Dst indices (STENNING, 1990).

The magnitude of Dst, and whether it is positive or negative, depends upon the proper subtraction of the main field from the H-component values for the representative stations. The earth's main field has a nonlinear secular change. The Dst provider must decide whether to accept the accuracy of those modeling this field about every five years for charting purposes or to determine the level uniquely for the contributing observatories. In addition, there is a need to remove the annual and semiannual changes in the main field (CAMPBELL, 1980) that respond to the magnetospheric distortions by a seasonal shift in the solar wind direction and remove the semiannual equinoxial period interactions (BOLLER and STOLOV, 1970).

Corrections for the frequency dependent earth-induced contribution to the H fields that compose Dst have been traditionally ignored by the index preparers. Recent advances in the understanding of the earth conductivity profiles (cf., CAMPBELL and SCHIFFMACHER, 1988) could be used to remove the internal contributions to the field observations and improve the external ring current representation.

The H-component direction, though generally pointing toward the geomagnetic pole, is greatly influenced by local crustal magnetic anomalies. This means that the assumption of a geomagnetic axial alignment of the index that comprises the H components is in difficulty. A better index should rotate the records to the appropriate dipole field axial alignment, using more than the local H component.

5. Concluding Remarks

The magnetospheric ring current, because of its stability, external location, and greater than one-day duration can provide a source for the determination of deep earth mantle conductivity. However, for effective use of this solar-terrestrial disturbance event, the researcher should understand the complicated magnetospheric current behavior sufficiently to select the proper geomagnetic disturbances, utilize only selective times in the events, obtain a proper distribution of observatory records, and remove competitive processes from the data (cf., BANKS, 1981). The assumption that the Dst is a "ring current index" overlooks the present knowledge of magnetospheric disturbance processes; in the present form, Dst is unsuited for use as an induction analysis source.

During the recovery phase of the geomagnetic storm the magnetosphere seems to be in the simplest of the disturbed conditions. It is during that period that the ring current field can best be used as a source for global induction analysis. Also, to enhance the purity of the records, I would recommend the following eliminations from the global observatory data set for the geomagnetic storm day: 1. High latitude sites poleward of the SAR arc boundary; 2. Equatorial locations within 600 km of the dip equator; and 3. Mid-latitude sites near the anticipated Sq current focus location for the analysis day. Before proceeding with the spherical harmonic analysis, field values for global regions devoid of observatories can be modeled by reasonable extrapolations from the remaining observatories.

REFERENCES

BANKS, R. J. (1969), *Geomagnetic Variations and Electrical Conductivity of the Upper Mantle*, Geophys. J. Roy. Astr. Soc. *17*, 457–487.

BANKS, R. J. (1972), *The Overall Conductivity Distribution of the Earth*, J. Geomag. Geoelectr. *24*, 337–351.

BANKS, R. J. (1981), *Strategies for Improved Global Electromagnetic Response Estimates*, J. Geomag. Geoelectr. *33*, 569–585.

BOLLER, B. R., and STOLOV, H. L. (1970), *Kelvin–Helmholtz Instability and the Semiannual Variation of Geomagnetic Activity*, J. Geophys. Res. *75*, 6073–6084.

CAMPBELL, W. H. (1980), *Secular, Annual, and Semiannual Changes in the Base-line Level of the Earth's Magnetic Field at North American Locations*, J. Geophys. Res. *85*, 6557–6571.

CAMPBELL, W. H., and SCHIFFMACHER, E. R. (1988), *Upper Mantle Conductivity for Seven Subcontinental Regions of the Earth*, J. Geomag. Geoelectr. *40*, 1387–1406.

CAMPBELL, W. H. *The regular geomagnetic-field variations during quiet solar conditions*, Chap. 6, In *Geomagnetism*, Vol. 3 (Jacobs, J. A., ed.) (Academic Press, London 1989) pp. 385–460.

CANTWELL, T. (1960), *Detection and Analysis of Low-frequency Magnetotelluric Signals*, Ph.D. thesis, Mass. Inst. Tech., 170 pp.

CHAPMAN, S. (1919), *The Solar and Lunar Diurnal Variation of the Earth's Magnetism*, Phil. Trans. Roy. Soc. *A218*, 1–118.

CHAPMAN, S., and WHITEHEAD, T. T. (1922), *The Influence of Electrically Conducting Material within the Earth as Inferred from Terrestrial Magnetic Variations*, Trans. Cambridge Phil. Soc. *22*, 463–482.

CHAPMAN, S. (1927), *On Certain Average Characteristics of World-wide Magnetic Disturbance*, Proc. Roy. Soc. London *A115*, 242–267.

CHAPMAN, S., and PRICE, A. T. (1930), *The Electric and Magnetic State of the Interior of the Earth as Inferred from Terrestrial Magnetic Variations*, Phil. Trans. Roy. Soc. London *A229*, 427–460.

CHAPMAN, S., and BARTELS, J., *Geomagnetism* (Oxford University Press, London 1940), 1049 pp.

DMITRIEV, V. I., ROTANOVA, N. M., ZAKHAROVA, O. K., and BALYKINA, O. N. (1977), *Geoelectric and Geothermal Interpretation of the Results of Deep Magnetic-variation Sounding*, Geomag. Aeron. *17*, 210–213 (English ed.).

DZIEWONSKI, A. M., and ANDERSON, D. L. (1981), *Preliminary Reference Earth Model*, Phys. Earth Planet. Int. *25*, 279–356.

FAINBERG, E. B., and ROTANOVA, N. M. (1974) *Distribution of Electrical Conductivity and Temperature in the Interior of the Earth According to Deep Electromagnetic Soundings*, Geomag. Aeron. *14*, 603–607 (English ed.).

FILLOUX, J. H. (1980), *Magnetotelluric Soundings over the Northeast Pacific May Reveal Spatial Dependence of Depth and Conductance of the Asthenosphere*, Earth and Planet. Sci. Lett. *46*, 244–252.

HOFFMAN, R. A., and BRACKEN, P. A. (1967), *Higher-order Ring Currents and Particle Energy Storage in the Magnetosphere*, J. Geophys. Res. *72*, 6039–6049.

IIJIMA, T., and POTEMRA, T. A. (1976), *Field-aligned Currents in the Dayside Cusp Observed by TRIAD*, J. Geophys. Res. *81*, 5971–5979.

IIJIMA, T., POTEMRA, T. A., and ZANETTI, L. J. (1990), *Large-scale Characteristics of Magnetospheric Equatorial Currents*, J. Geophys. Res. *95*, 991–999.

ISIKARA, A. M. (1980), *Long-period Variations of the Geomagnetic Field and Inferences about the Deep Electric Conductivity*, J. Geomag. Geoelectr. *32*, Suppl. 1, 155–157.

KOVTUN, A. A., and POROKHOVA, L. N. (1980), *Deep Conductivity Distribution on Russian Platform from the Results of Combined Magnetotelluric and Global Magnetovariational Data Interpretation*, J. Geomag. Geoelectr. *32*, Suppl. 1, 105–113.

LAHIRI, B. N., and PRICE, A. T. (1939), *Electromagnetic Induction in Nonuniform Conductors, and the Determination of the Conductivity of the Earth from Terrestrial Magnetic Variations*, Phil. Trans. Roy. Soc. London *A237*, 509–540.

LUI, A. T. Y., McENTIRE, R. W., and KRIMIGIS, S. M. (1987), *Evolution of the Ring Current during Two Geomagnetic Storms*, J. Geophys. Res. *92*, 7459–7470.

MATSUSHITA, S., TARPLEY, J. D., and CAMPBELL, W. H. (1973), *IMF Sector Structure Effects on the Quiet Geomagnetic Field*, Radio Sci. *8*, 963–972.

MAYAUD, P. N., *Derivation, Meaning, and Use of Geomagnetic Indices* (Amer. Geophys. Union Monograph, Washington, D.C. 1980), 153 pp.

McDonald, K. (1957), *Penetration of the Geomagnetic Secular Field Through a Mantle with Variable Conductivity*, J. Geophys. Res. *62*, 117–141.

Moos, N. A. F., *Magnetic observations made at the government observatory Bombay for the period 1846 to 1905 and their discussion, Part II: The phenomenon and its discussion*, In *Colaba Magnetic Data* (Bombay Observatory Publication 1910).

Price, A. T. (1973), *The Theory of Geomagnetic Induction*, Phys. Earth Planet. Int. *7*, 227–233.

Rangarajan, G. K., *Indices of geomagnetic activity*, Chap. 5, In *Geomagnetism*, Vol. 3 (Jacobs, J. A., ed.), (Academic Press, London 1989) pp. 323–384.

Rastogi, R. G., *The equatorial electrojet: Magnetic and ionospheric effects*, Chap. 3, In *Geomagnetism*, Vol. 3 (Jacobs, J. A., ed) (Academic Press, London 1990) pp. 461–525.

Rees, M. H., and Roble, R. G. (1975), *Observation and Theory of the Formation of Stable Auroral Red Arcs*, Rev. Geophys. Space Phys. *13*, 201–242.

Rikitake, T. (1950), *Electromagnetic Induction within the Earth and its Relation to the Electrical State of the Earth's Interior, Part II*, Tokyo Univ. Bull. Earthquake Res. Inst. *28*, 263–283.

Rikitake, T., *Electromagnetism and the Earth's Interior* (Elsevier Pub. Co., Amsterdam 1966), Chap. 15, pp. 221–230.

Roelof, E. C., and Williams, D. J. (1988), *The Terrestrial Ring Current: From in situ Measurements to Global Images Using Energetic Neutral Atoms*, Johns Hopkins APL Tech. Dig. *9* (2), 1–19.

Rokityansky, I. I., *Geolectromagnetic Investigation of the Earth's Crust and Mantle* (Springer Verlag, Berlin 1982), 381 pp.

Schuster, A. (1890), *The Diurnal Variations of Terrestrial Magnetism* Phil. Trans. Roy. Soc. *A130*, 467–512.

Stenning, R. J. (1990), *A Lunar Tide in the Dst Index*, J. Geomag. Geoelectr. *42*, 11–17.

Sugiura, M. (1964), *Hourly Values of the Equatorial Dst for IGY*, Ann. Int. Geophys. Year *35*, 9–51.

Tarpley, J. D. (1973), *Seasonal Movement of the Sq Current Foci and Related Effects in the Equatorial Electrojet*, J. Atoms. Terr. Phys. *35*, 1063–1071.

Tozer, D. C. (1970), *Temperature, Conductivity, Composition and Heat Flow*, J. Geomag. Geoelectr. *22*, 35–51.

Williams, D. J., and Sugiura, M. (1985), *The MPTE Charge Composite Explorer and the 4–7 September 1984 Geomagnetic Storm*, Geophys. Res. Lett. *12*, 305–308.

Yukutake, T. (1965), *The Solar Cycle Contribution to Secular Change in the Geomagnetic Fields*, J. Geomag. Geoelectr. *17*, 287–309.

(Received August 10, 1990, revised/accepted August 21, 1990)

PAGEOPH, Vol. 134, No. 4 (1990)

0033–4553/90/040559–16$1.50 + 0.20/0

The Conductosphere Depth at Equatorial Latitudes as Determined from Geomagnetic Daily Variations

S. Duhau[1] and A. Favetto[1]

Abstract — The structure of the conductosphere at equatorial latitudes is investigated by analyzing geomagnetic daily variations (GDV) data. To do so the magnetic field induced in a multilayered model by an external source of finite size that reproduces the external part of these variations is computed; latitudinal profiles of the total field produced at ground by the assumed external source are then found and compared with the measured GDV data. Results are applied to equatorial zones in Peru, Africa and India.

Key words: Conductosphere, upper mantle, GDV, earth conductivity.

1. Introduction

Electromagnetic sounding methods give generally less resolution than others, such as seismic, but they provide some complementary information (see e.g. ROBERTS, 1983).

GDV comprise such a broad band of frequencies that the study of the earth's electromagnetic response to that field allows us to obtain electrical conductivity from shallower zones to very deep structures, due to the depth of penetration of these variations being frequency dependent.

With regard to geomagnetic field variations due to natural external sources, there are basically two methods for investigating the conductivity of the earth from its surface to upper mantle depths. These are (1) the magnetotelluric (MT), where two horizontal, mutually orthogonal magnetic and electric fields are recorded over a range of frequencies, and (2) the geomagnetic depth sounding (GDS) method, where the three components of the magnetic field are recorded. GDV are the external source field variations that contain frequencies low enough to provide additional and complementary information about the deeper structures, in particular, when the field induced within the earth is obtained by a spherical harmonic analysis of the GDV recorded at geomagnetic observatories worldwide, and is interpreted through a spherical layered model of the earth; then a sharp increase in

[1] Departamento de Física, Universidad de Buenos Aires, 1428 Ciudad Universitaria Pab. I, Buenos Aires, Argentina.

the conductivity is found at a depth of about 650 km, where it is about 10^4 times larger than others at crustal depths (LAHIRI and PRICE, 1939). More recent global results based on the spectral analysis of average daily values of the geomagnetic components, recorded also worldwide, indicate the existence of two main layers where the conductivity increases sharply; the first at a depth of approximately 500 km and the second at approximately 1200 km. Their values are between 0.1 to 1 Mho/m and 10 to 100 Mho/m respectively (see e.g., PĚČOVÁ et al., 1987).

GDV penetrate deep enough that from their analysis it is possible to estimate the depth at which that second highly conducting zone that we called conducto-sphere, begins (see e.g., DUHAU and ROMANELLI, 1979). At daily frequencies the field induced within the earth depends on the size of the source, contrary to what happens at higher frequencies where the source may be considered uniform (see e.g., PRICE, 1967). This dependence becomes more important at equatorial lati-tudes, where the east-west electrojet flows at ionosphere heights in a band as narrow as 800 km. Since the size of the source is not taken into account in MT and GDS sounding these methods can badly bias the estimate of the conductivity of the earth at upper-mantle depths. To solve this problem at equatorial latitudes, it is necessary to know the spatial distribution of the source. The way to determine it from the GDV has been carefully discussed by DUHAU and OSELLA (1982), who, comparing in situ data of the ionospheric current with the external current inferred from the GDV showed that: i) 90% of the external source of the GDV is flowing in the E region of the ionosphere, ii) the external source may be considered flowing in a thin layer in order to compute the GDV from it and, finally, iii) the latitudinal variation of its intensity may accurately be inferred from the GDV.

To analyze the GDV at equatorial latitudes, every method (see e.g., FORBUSH and CASAVERDE, 1961; FAMBITACOYE and MAYAUD, 1976; ONWUMECHILLI, 1967) starts by separating the field variations into its external and internal contribu-tions, that is to say, produced by the external currents and the currents induced within the earth, respectively. The external current is then inferred from the external contribution thus separated and the internal contribution is compared with the field induced by that current system within a proposed earth's model (see e.g., FAMBITA-COYE and MAYAUD, 1976; OSELLA and DUHAU, 1985). Improvements to the method introduced by DUHAU and OSELLA (1982) allow the performance of the first step of the analysis of the GDV without making any a priori hypothesis about the magnitude of the field induced within the earth. As a result, the precision required to deduce the depth at which the conductosphere begins is greatly increased in areas where not only the value of this depth is suspected to differ from its average value, but also where lateral discontinuities may exist (DUHAU et al., 1982 and DUHAU and OSELLA, 1983).

Using a simple analytical model representing the particular features of the equatorial current system, the above-mentioned analysis has been applied to different zones near the dip equator in Peru (OSELLA and DUHAU, 1983), Nigeria

(DUHAU and OSELLA, 1983), Central Africa (DUHAU and OSELLA, 1984) and India (FAVETTO et al., 1990) assuming a simple earth's model consisting of a nonconducting layer above a perfectly conducting half-space. It was found that not only the depth, p, at which the conductosphere begins is different in Peru from that in Africa but also that a significant gap in p between the North and the South of the dip equator exists in Africa and Peru.

Since in all the above-mentioned papers, the conductivity of the layers above the conductosphere has been assumed to be negligible, to evaluate the possible error that this assumption could introduce on the determination of the conductosphere depth DUHAU and FAVETTO (1990), again used, a two-layered model, but now, the upper layer was assumed to have a finite conductivity to represent the structure above the conductosphere. The field induced by an harmonic source that could represent the average global shape of the external source, has been computed on a wide range of conductivities which comprises the values given by MT and GDS methods. Also it was found that the influence of the upper layers as expected, increases with its conductivity and deepness. It is disregarded in Peru where the conductosphere depth is less then 400 km (DUHAU and OSELLA, 1983) and gives an appreciable contribution at zones, such as the South of the dip equator in Africa or the North in India where the conductosphere begins deeper than 1000 km (DUHAU and OSELLA, 1984; FAVETTO et al., 1990).

In the present paper we further investigate the influence of the finite conductivity of the upper layers by determining the conductosphere depth. To do so, the field induced by the equatorial current system, determined from the GDV, is calculated assuming a multilayered model. To build up that model, values of the upper layer conductivities and thickness available from MT and GDS data in Africa (see e.g., RITZ, 1983; RITZ and ROBINEAU, 1986) are used. To find the field induced by the external current within this more realistic multilayered earth model, the theory for a two-layered model introduced by PRICE (1967) was generalized.

To estimate the effect of the upper layers conductivities on the determination of p, we have centered the discussion on Central African data, not only because according to previous results it is appreciable within that zone but also, as we will discuss, it allows extention of the results to more general conclusions due to the fact that this zone does not present significant localized anomalous conductivity features within the upper layers, showing the typical underground conductivities that are expected to be found associated with its local tectonic features (see e.g., ROBERTS, 1983).

2. The GDV at Equatorial Latitudes

2.1 The External and Internal Fields

The external and internal contribution of the geomagnetic field variations may be separated by using the integral transform method for bi-dimensional fields proposed by SIEBERT and KERTZ (1957) in any case in which a profile exists called "allowed profile" where these variations fulfill this condition. Since the external current mainly flows from east to west at the equator, any north-south profile may be chosen as allowed, provided a significant east-west variation in the internal conductivity does not exist in the zone. This could be checked by observing whether the east-west derivative of the field declination, D, is negligible in the Maxwell equation that expresses the divergence less of the geomagnetic field at the ground. In that case the horizontal, H, and vertical, Z, components of the GDV fulfill the following relationships:

$$H_\theta = KZ_\theta \tag{1a}$$

$$Z_\theta = -KH_\theta \tag{1b}$$

$$H_i = -KZ_i \tag{1c}$$

$$Z_i = KH_i \tag{1d}$$

where the subscripts θ and i mean the external and internal contribution respectively and K is the Kertz operator defined by the Hilbert transform of the field components, $f(x)$, as

$$Kf(x) = \frac{1}{\pi} \int_{-\infty}^{\infty} \frac{f(x')}{(x-x')} dx'$$

where the integration is made along the allowed profile.

Using equations (1a) through (1d) the external and internal parts of the field may be expressed as a function of the total field variations, $H = H_\theta + H_i$ and $Z = Z_\theta + Z_i$, as follows

$$H_\theta = \tfrac{1}{2}[H + KZ] \tag{2a}$$

$$H_i = \tfrac{1}{2}[H - KZ] \tag{2b}$$

$$Z_\theta = \tfrac{1}{2}[Z - KH] \tag{2c}$$

$$Z_i = \tfrac{1}{2}[Z + KH]. \tag{2d}$$

If the total field components do not tend to zero at the end of the finite interval within which they are measured, to compute the Kertz operator, it is necessary to provide a continuation of them beyond the limits of that interval. This poses a special problem at the equator.

Different approaches have been made to solve this problem. Several authors first separate the H and Z components of the total field into a localized, "incremental" field and an extended, "planetary" field. They applied eq. (2) only to separate the incremental field into its internal and external parts. To separate the planetary field, they assumed its internal part as forty percent of its external one, a percentage that is deduced from a weighted average of the first three coefficients of a spherical harmonic of the GDV at planetary scale (see e.g., FORBUSH and CASAVERDE, 1961; FAMBITACOYE and MAYAUD, 1976). Since this procedure gives only the value of the internal contribution averaged over the entire earth corresponding to an approximated value for p of about 650 km (see e.g., OSELLA and DUHAU, 1985), then, it introduces an error when the local value of p is distant from this average, which is the rule instead of the exception (see e.g., ROBERTS, 1983). For the purpose of using GDV to find local variations in the value of p, DUHAU and OSELLA (1982) have applied the Siebert and Kertz method without making any *a priori* assumption about the value of the induced planetary field. To do so, these authors applied the Kertz operator only to the Z component by providing an adequate continuation of it beyond the measured interval, then according to equation (2), only the H component was separated using that method. This has been done so because the application of the Kertz operator is more reliable when the field components show low values at the end of the measured interval (SIEBERT and KERTZ, 1957).

2.2 The Current System Model

Once the external horizontal field, H_θ, is obtained following the above-cited procedure, the current system may be deduced, assuming that it is flowing in a thin layer concentric to the earth. Considering that the magnetic field is given by an integral expression of this current through the Biot and Savart law, no fine detail may be found from that field. Consequently it is convenient to assume that the current system is given by a simple analytical expression. Also, for convenience, at equatorial latitudes this system may be represented as the superposition of a planetary and an incremental part. Of course, this is only a representation and has no physical meaning, thus when applying the results, the total current and total geomagnetic field should be computed first.

One simple and useful expression to represent the current system is:

$$\mathbf{J}_\theta^j(x, z) = J_0^j\left(1 - \frac{x^2}{D^2}\right)\delta(z + h)\hat{y} \qquad -D < x < D$$

$$\mathbf{J}_\theta^j(x, z) = 0 \qquad\qquad\qquad x > D \quad \text{and} \quad x < D \qquad (3)$$

for the incremental current system (see e.g., CHAPMAN, 1951; DUHAU and OSELLA, 1983) where x and z are the eastward and downward coordinates, respectively and

$$\mathbf{J}_\theta^P(x, z) = (J_v + J_c)\delta(z + h)\hat{y} \qquad (4a)$$

$$J_v(x, z) = \frac{2}{\mu_0}[A\, e^{-kz}\cos(k(x - x_0)] \qquad (4b)$$

$$J_c(x, z) = \frac{2}{\mu_0}C \qquad (4c)$$

for the planetary one (OSELLA and DUHAU, 1983), where J_0^i is the amplitude, $h = 107$ km is the height at which it circulates (see e.g. MAYNARD, 1967), D is the width of the electrojet, δ is the Dirac delta function and μ_0 is the magnetic susceptibility of vacuum.

The parameters A, C, x_0, D and J_0^i may be found by adjusting the field produced at ground by the current system given in eqs. (3) and (4) to the external field obtained by separating the GDV.

3. Field Induced at Ground by the GDV Current System

3.1 Theory

3.1.1 Electromagnetic response of a layered model to any electromagnetic source. To take into account the finite conductivities of the upper layers we generalize to the multilayered case the results obtained for a two-layered model given by PRICE (1950). Since the basic physical hypothesis is the same in both cases, a brief summary of both is given below for completeness.

A plane-layered conductor occupying the half-space, $z > 0$, is considered and the magnetic field on the surface ($z = 0$) is found. Within the earth, the conductivity, σ, is assumed to vary with depth considerably more than the dielectric constant, and the magnetic susceptibility, μ; in order to simplify mathematical calculations they are taken to be constants.

From Maxwell's theory and the media constitutive equations, a differential equation system is obtained which together with the boundary conditions permits solution of the problem of a varying magnetic field of any distribution having its source in the region $z < -h$.

Since the period, τ, of the electromagnetic waves is large compared with the time needed to travel across the region, the displacement current is disregarded in the Ampere law. It is also assumed that the electric field, E, induced by any magnetic field of external origin, can be built from elementary solutions of the form:

$$\mathbf{E} = G(z, t)\mathbf{F}(x, y) \tag{5}$$

applying the Maxwell equation in the special form that results from the above approximations, it is found that $\mathbf{F}(x, y)$ may be written as:

$$\mathbf{F}(x, y) = \left(\frac{\partial P}{\partial y}(x, y), \ -\frac{\partial P}{\partial x}(x, y), 0 \right) \tag{6}$$

and $G(z, t)$ satisfies

$$\frac{\partial^2}{\partial z^2} G(z, t) = \left\{ k^2 + \mu\sigma \frac{\partial}{\partial t} \right\} G(z, t) \tag{7}$$

with

$$\frac{\partial^2}{\partial x^2} P(x, y) + \frac{\partial^2}{\partial y^2} P(x, y) + k^2 P(x, y) = 0 \qquad (8)$$

where k is the separation constant.

The solution of (7) is

$$G(z, t) = G_0 e^{-kz} + G_0' e^{kz} \quad \text{outside the conductor.}$$

If we take k as real and positive, the term involving e^{-kz} corresponds to a magnetic-induced field while the term involving e^{kz} corresponds to the field due to the induced currents. If we assume the induced field is periodic, with frequency ω and that the conductivity is uniform for each layer, then the solution of (7) in the conductor takes a generalized form for each layer

$$G_i(z, t) = G_i e^{-\theta_i z} + G_i' e^{\theta_i z}$$

where

$$\theta_i^2 = i\sigma_i \mu \omega + k^2.$$

The magnetic induction, $\mathbf{B}$, may be obtained from the electric field by applying Faraday's law. After replacing $F(x, y)$ given by eq. (6) in the expression of the electric field (eq. (5)), it is found that for the magnetic induction, $\mathbf{B}$, the following holds

$$-\frac{\partial}{\partial t} \mathbf{B} = \left(\frac{\partial G}{\partial z} \frac{\partial P}{\partial x}, \ \frac{\partial G}{\partial z} \frac{\partial P}{\partial y}, \ k^2 GP \right) \qquad (9)$$

where $\mathbf{B}$ is bidimensional, H and Z denote its horizontal and vertical components respectively.

3.1.2 Field induced within a layered model by a unidirectional source — Harmonic source. The field produced by an harmonic source flowing in y direction, is given by the solution, $P(x, y)$, of eq. (8)

$$P(x, y) = \frac{1}{k} \sin(k(x - x_0))$$

which according to eq. (9) leads to the following expression

$$H_\theta = A \cos(k(x - x_0))$$
$$Z_\theta = -A \sin(k(x - x_0)) \qquad (10)$$

for the field components at ground.

When the boundary conditions are applied to the electric and magnetic fields given by equations (5) and (9) at all surfaces separating different media, it is possible to find the field induced at ground by the external one (eq. (10)) as a function of the following

$$H_i = \alpha H_\theta$$

$$Z_i = -\alpha Z_\theta. \tag{11}$$

The algebraic expression of α corresponding to an n-layered earth model, is given by

$$\alpha = \frac{(k - \theta_1) + M_1(k + \theta_1)}{(k + \theta_1) + M_1(k - \theta_1)} \tag{12}$$

with the recurrence formula

$$M_i = \frac{(\theta_i - \theta_{i+1}) e^{-2\theta_i + 1P_i} + M_{i+1}(\theta_i + \theta_{i+1})}{(\theta_i + \theta_{i+1}) e^{-2\theta_i P_i} + M_{i+1}(\theta_i - \theta_{i+1})} e^{-2\theta_i P_i}$$

for $i < n$,

$$M_n = -e^{-2p_n \theta_n}$$

where p_n is the depth at which the conductosphere begins and the subscript i denotes the i-th layer.

—General case. To analyze the field induced by a source that shows latitudinal variations, we considered it as a continuous distribution of current lines flowing in the y direction with intensity $J(x')$. Then, the expression of the external field at the earth's surface is (PRICE, 1950):

$$H_\theta = \frac{\mu_0}{2\pi} \int_{-\infty}^{\infty} J(x') \left\{ \int_0^{\infty} e^{-kh} \cos(k(x - x')) \, dk \right\} dx'$$

$$Z_\theta = -\frac{\mu_0}{2\pi} \int_{-\infty}^{\infty} J(x') \left\{ \int_0^{\infty} e^{-kh} \sin(k(x - x') \, dx' \tag{13}$$

the field induced at ground for each wave number k, is given by eq. (11), then

$$H_i = \frac{\mu_0}{2\pi} \int_{-\infty}^{\infty} J(x') \left\{ \int_0^{\infty} \alpha(k) e^{-kh} \cos(k(x - x')) \, dk \right\} dx'$$

$$Z_i = \frac{\mu_0}{2\pi} \int_{-\infty}^{\infty} J(x') \left\{ \int_0^{\infty} \alpha(k) e^{-kh} \sin(k(x - x')) \, dk \right\} dx' \tag{14}$$

where α is given by eq. (12).

3.2 Application

The ionospheric current system that produces the GDV at equatorial latitudes flows east-westerly and may be represented by the sum of an incremental (eq. (3)) and a planetary (eq. (4)) part. Also the latter is a sum of a variable part (eq. (4b)) and other constant one (eq. (4c)). The Maxwell equations being linear we may solve separately the problem for each of these three parts and then find the final solution to their superposition.

The planetary external field computed by applying Biot and Savart law to the current system, given by eq. (4), is

$$H_\theta^P = H_{\theta v}^P + C$$
$$Z_\theta^P = Z_{\theta v}^P \tag{15}$$

where

$$H_{\theta v}^P = A \cos(k(x - x_0))$$
$$Z_{\theta v}^P = -A \sin(k(x - x_0)). \tag{16}$$

The internal part of the harmonic and constant contributions to the planetary field may be computed from eqs. (11) and (14) with $J(x') = 2/\mu_0 \, C$ and summing both, the planetary induced field is found by

$$H_i^P = \alpha H_{\theta v}^P + C$$
$$Z_i^P = -\alpha Z_\theta^P. \tag{17}$$

Finally, external and internal parts of the incremental field may be computed from eqs. (13) and (14), respectively after introducing the expression of $J(x')$ given by eq. (3).

4. Results

4.1 The GDV Data

GDV data have been recorded at equatorial latitudes, throughout most north-south profiles over the different continents. In Africa by FAMBITACOYE (1973) at observatories widely distributed around the dip equator and by ONWUMECHILLI (1967) in a more dense chain but covering a narrower interval of latitudes. In Peru, by FORBUSH and CASAVERDE (1961) and in India by the Indian Institute of Geomagnetism (1982).

4.2 Africa

Figure 1 shows the external and internal horizontal field separated by DUHAU and OSELLA (1984) from the field measured at ground in Central Africa by FAMBITACOYE and MAYAUD (1976). The parameters to describe the current system are: A, C, x_0 and J_0^i as defined in eqs. (3) and (4) and obtained from the external

S. Duhau and A. Favetto

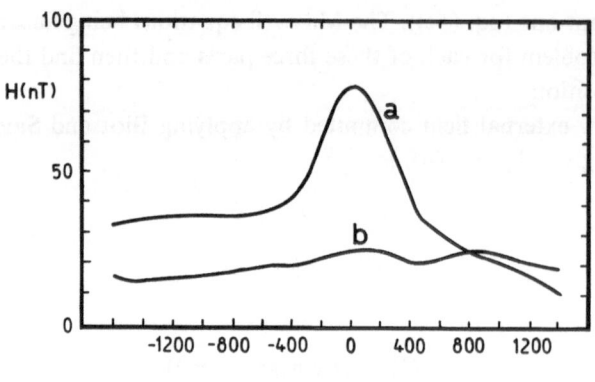

DISTANCE TO THE DIP EQUATOR (km)

Figure 1

The external (a) and internal (b) parts of the horizontal component of the geomagnetic daily variations as separated from the total field by DUHAU and OSELLA (1984).

horizontal field (see curve (a) in Figure 1). Also used in the present work are the ones which these authors have obtained in this region (see Table 1).

Magnetotelluric and GDS data recorded by RITZ (1983) and RITZ and ROBINEAU (1986) in the African craton and basin, respectively, provide a conductivity distribution close to a depth of 500 km which is different below each of the tectonic features present in the zone. Noticeably a layer with a conductivity of 0.1 Mho/m has been inferred to be placed at a 300 km depth below the basin and at a 500 km depth below the craton. This fact indicates the existence of a deep lateral discontinuity. This discontinuity still seems to exist very deeply since DUHAU and OSELLA (1984) found, through analysis of the GDV and interpretation of the internal field with a model which neglects the upper layers conductivity, that the conductosphere begins at a depth of 500 km below the basin and 1000 km below the craton.

In order to analyze the GDV, taking into account the upper layers conductivity, the procedure described in previous sections was followed. The external and

Table 1

Current system parameters determined by DUHAU and OSELLA (1984)

Planetary current	Incremental current
$k = 7.6 \ 10^{-4} \ \mathrm{km}^{-1}$	$D = 350 \ \mathrm{km}$
$A = 19 \ \mathrm{nT}$	$x_j = 40 \ \mathrm{km}$
$C = 15 \ \mathrm{nT}$	$J_0^j = 144.04 \ \mathrm{nT}/\mu_0$
$x_0 = -700 \ \mathrm{km}$	

internal parts of the field produced by the planetary (eq. (4)) and incremental (eq. (3)) current system at ground have been computed following the steps presented in Section 3.2 and summing both contributions, a latitudinal profile of the total field has thus been obtained and compared with the data. Two different models representing the structure underlying the basin and craton tectonic features have been used in the present paper. To build these models, the upper layers structure has been taken from the 2D profile proposed by RITZ (1983) and RITZ and ROBINEAU (1986) assuming that the higher values found to be representative of each layer and a layer just below were added to represent the conductosphere as was found by DUHAU and OSELLA (1984) (see Figure 2). When we refer to the basin or craton model we mean those conducting structures which are located at the North and South of the dip equator, respectively. Certainly as the structure below the craton and the basin zones may vary from place to place, the assumed models do not represent the actual situation at any place in particular; but, since the larger measured conductivities of each of the layers above the conductosphere were considered to build up, the contribution of these layers to the internal part of the GDV, as compared with that of the conductosphere, may be safely evaluated by assuming this model describes the structure.

Figure 3 shows the result when the basin and craton models are considered. Notice that for both components, H and Z, the craton model fits well to the South and the basin model to the North. Then a strong North-South variation on the conductivity structure exists within a narrow interval around the dip equator. Certainly, this discontinuity is well apparent, in that the Z component is the reason that none of the models fit well Z data within the interval (-500 km, 500 km) around the dip equator. The fact that Z data points beyond that interval are fitted

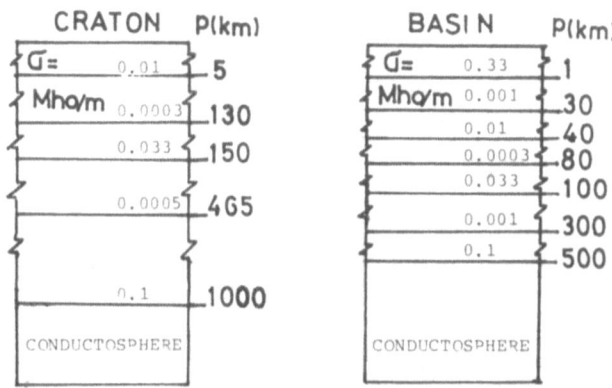

Figure 2

Multilayered model of the earth used to analyze the GDV in Africa (the depth and conductivity of the upper layers as given by RITZ (1983) and RITZ and ROBINEAU (1986) and the depth at which the conductosphere begins as given by DUHAU and OSELLA (1984)).

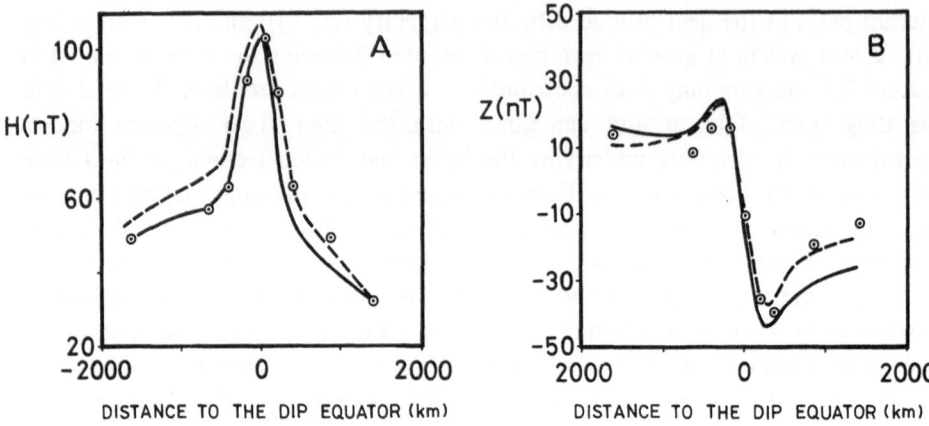

Figure 3

Horizontal (3A) and vertical (3B) components of the GDV as a function of the distance to the dip equator. 1) Data points (⊙) (FAMBITACOYE, 1973) and 2) profile calculated as the sum of the external field produced at ground by the current system given in Section 2.2 and the field induced by it on the conducting structure schemed in Figure 2, for the craton (——) and for the basin (– – –).

quite well by the corresponding model indicates that the strong variation on the earth structure occurs in an interval of latitude equal to, or smaller than, this.

We may conclude that there is a discontinuity on the conductivity between basin and craton, not only at the depth at which the highly conducting layer ($\sigma = 0.1$ Mho/m) placed immediately above the conductosphere begins, but also at the depth at which this deeper and even more conductive layer, the conductosphere, begins. This discontinuity takes place in a very narrow band of latitudes around the dip equator, whose extension is equal to or smaller than the size of the discontinuity itself.

It should be noted that the results do not change if the resistivities of the upper layers are neglected and only the conductosphere and the highly conducting layer immediately above are considered. Thereafter, a two-layered model is enough to interpret geomagnetic variations at such a low frequency as that corresponding to the GDV (≈ 1 day).

To illustrate further the significance of these two layers in Figures 4 and 5, the results for two alternative models are shown, one in which only the conductosphere is taken into account and the other in which only a half-space of 0.1 Mho/m is considered. We observe that both models give equivalent results for craton and then cannot be distinguished from the GDV, while for the basin the 0.1 Mho/m layer may be neglected. Since this layer is placed above the conductosphere it has a very small effect on GDV as compared to the one given by the conductosphere itself.

We conclude that below the basin, conductosphere depth is around 500 km and below craton the 0.1 Mho/m layer is thick enough to shield the conductosphere at

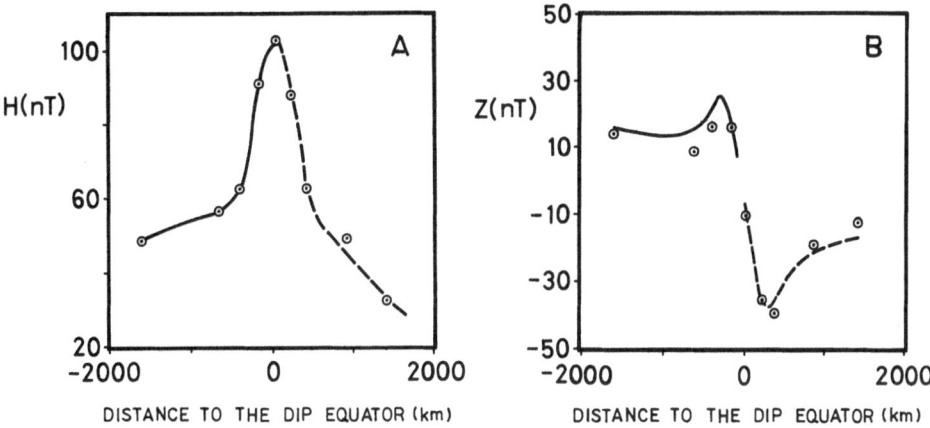

Figure 4

Horizontal (4A) and vertical (4B) components of the GDV as a function of the distance to the dip equator. 1) Data points (⊙), (FAMBITACOYE, 1973) and 2) *Idem* Figure 3 but now the induced field is inferred neglecting the conductivity of all the upper layers and assuming that the conductosphere (100 Mho/m) begins at: 1000 km, for the craton (——), and 500 km, for the basin (– – –).

GDV frequencies. This could be located even deeper than the value of 1000 km previously reported by DUHAU and OSELLA (1984).

Now, a question remains to be answered: Is this strong, deep discontinuity in the earth's structure completely associated with the crust and surface tectonics or do we have another condition? Data from other places offer some insight on this point upon which we will comment in the following sections.

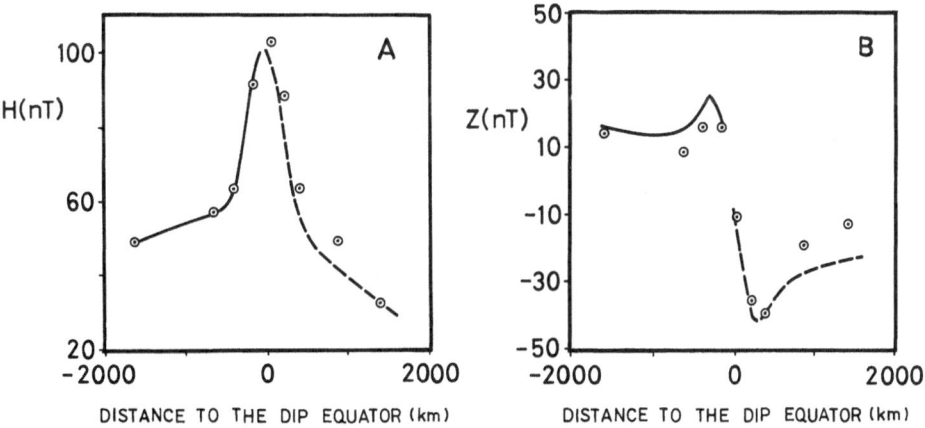

Figure 5

Horizontal (5A) and vertical (5B) components of the GDV as a function of the distance to the dip equator. 1) Data points (⊙), (FAMBITACOYE, 1973) and 2) *Idem* Figure 4 but now the induced field is inferred, assuming a half-space of conductivity 0.1 Mho/m at a depth of: 465 km, for the craton (——), and 300 km, for the basin (– – –).

4.3 India

The result of the GDV analysis in Africa (Section 4.2) can be extended to India. There, data have been mostly recorded in the North because the coast is very close to the dip equator to the South.

FAVETTO et al. (1990) have inferred the depth at which the conductosphere begins, evaluating the induction effect through a model of the earth where the upper layers' conductivity is neglected. The result was that the depth at which the conductosphere begins is around 1000 km under the Indian Shield. By comparing this result with the one obtained at the South of the dip equator in Africa, we can conclude that in both cases, since in India MT and GDV indicate similar values of conductivity for the upper structure, including the 0.1 Mho/m layer (see e.g., CHAMALAUN et al., 1987; RAMASWAMY et al., 1985), the conductosphere is also placed very deep there, even below 1000 km.

4.4 Peru

DUHAU and OSELLA (1983) estimated the conductosphere depth around the dip equator, neglecting all the upper conductivities. The values of p thus obtained are 150 km at the South of the dip equator and 450 km at the North. These values are low enough so that if we extend our analysis of Section 4.2 we can conclude that the result should not be modified by including in the model the conductivities of the upper layers. It is noteworthy that the gap found in p is significant enough to show the presence of a lateral discontinuity around the dip equator which, in this case, is not associated with a tectonic discontinuity, since the geomagnetometers chain there is located along the same tectonic feature.

4.5 Summary

In summary, below the Indian and African cratons the conductosphere is located deeper than 1000 km, below the African basin at a depth of 500 km and below the Tertiary folding mountains in Peru there is a lateral discontinuity in the upper mantle that is not correlated to any surface tectonic discontinuity, this depth being 150 km to the South of the dip equator and 450 km to the North.

4.6 A Remark About the Earth's Structure Below the Dip Equator

Notice that the dip equator in Africa follows the line defining the boundary between the craton and basin that is running along a deep discontinuity in the earth's structure. Additionally a deep discontinuity is found where the dip equator is located in Peru. This constitutes an initial insight of the possibility that the actual position of the dip equator could be associated, at less partially, with very deep discontinuities in the earth conductivity. More data around the dip equator should be obtained and analyzed in other places to confirm this early insight and to investigate further whether the same tectonic features are or are not correlated to the same deep structures.

5. Conclusions

The depth at which the conductosphere begins at equatorial latitudes is well defined from the GDV analysis when located at a depth above approximately 500 km, in which case the conductivity of the upper layers may be neglected and a minimum value for the conductosphere conductivity must be 10 Mho/m to provide a suitable representation of the GDV in these cases. Contrary to what occurs when the conductosphere is located very deep, (more than 1000 km) in which case the highly conducting layer (0.1 Mho/m) that is contiguously located and above it shields the conductosphere and only a lower limit of p may be inferred from the GDV, in actuality the conductosphere may be placed much deeper than this limit. Frequencies lower than the daily ones should be analyzed to find p more accurately in these cases. If we take into account that a great number of models describing the conductivity distribution are available from a layered spherical earth and many of those show that the conductivity increases to values of 0.1–1 Mho/m at about 500 km and 10–100 Mho/m at about 1200 km (see e.g., PÉČOVÁ et al., 1987), we conclude that our method allows us to find additional information about this last layer. The conductosphere from GDV data provides a determination of local variation of the depth at which this layer begins, which has proved to be, in many cases, well above or well below the average global value inferred from spherical models.

Acknowledgement

This work was partially supported by CONICET (Consejo de Investigaciones Científicas y Técnicas). The authors acknowledge helpful suggestions by one unknown referee.

REFERENCES

CHAMALAUN, F. H., PRASAD, S. N., LILLEY, F. E. M., SRIVASTAVA, B. J., SINGH, B. P., and ARORA, B. R. (1987), On the Interpretation of the Distinctive Pattern of Geomagnetic Induction Observed in Northwest India, Tectonophysics 140, 247.
CHAPMAN, S. (1951), The Equatorial Electrojet as Detected from the Abnormal Electric Current Distribution Above Huancayo, Peru, and Elsewhere, Ach. Meteorol. Geophys. Bioklimatol. A4, 368.
DUHAU, S., and FAVETTO, A. (1990), Estimacion de la contribucion a las variaciones geomagneticas diarias de los campos inducidos en las capas superiores de la tierra, Geoacta 17, 1.

DUHAU, S., and OSELLA, A. M. (1982), *A Correlation between Measured E-region Current and Geomagnetic Daily Variation at Equatorial Latitudes*, J. Geomag. Geoelectr. *34*, 213.

DUHAU, S., and OSELLA, A. M. (1983), *Depth of the Nonconducting Layer at the Nigerian Dip Equator* J. Geophys. Res. *88* (A7), 5523.

DUHAU, S., and OSELLA, A. M. (1984), *Depth of the Nonconducting Layer at Central Africa*, J. Geomag. Geoelectr. *36* (3), 113.

DUHAU, S., and ROMANELLI, L. (1979), *Electromagnetic Induction at South American Geomagnetic Equator as Determined from Measured Ionospheric Currents*, J. Geophys. Res. *84* (A5), 1849.

DUHAU, S., ROMANELLI, L., and HIRSCH, F. A. (1982), *Indication of Anomalous Conductivity at the Nigerian Dip Equator*, Plan. Space Sci. *30* (1), 97.

FAMBITACOYE, O. (1973), *Effects Induits par l'electrojet equatorial au centre de l'Afrique*, Ann. Geophys. *29*, 149.

FAMBITACOYE, O., and MAYAUD, R. N. (1976), *Equatorial Electrojet and Regular Daily Variations Sr: I. A. Determination of Equatorial Electrojet Parameters*, J. Atmosph. Terr. Phys. *38*, 1.

FAVETTO, A., OSELLA, A. M., and DUHAU, S. (1990), *Depth of the Conductosphere under the Indian Shield*, submitted for publication.

FORBUSH, S., and CASAVERDE, M. (1961), *Equatorial Electroject in Peru*, Carnegie Inst. Washington Pub. 620

LAHIRI, B. N., and PRICE, A. T. (1939), *Electromagnetic Induction in Nonuniform Conductors, and the Determination of the Conductivity of the Earth from Terrestrial Magnetic Variations*, Phil. Trans. Roy. Soc. London *A237*, 509.

MAYNARD, N. C. (1967), *Measurements of Ionospheric Currents off the Coast of Peru*, J. Geophys. Res. *72*, 1863.

ONWUMECHILLI, C. A., *Physics of Geomagnetic Phenomena*, Vol. 1 (Matsushita, B., and Campbell, C.) (Academic Press, New York 1967) 462 pp.

OSELLA, A. M., and DUHAU, S. (1983), *The Effect of the Depth of the Nonconducting Layer on the Induced Magnetic Field at the Peruvian Dip Equator*, J. Geomag. Geoelectr. *35*, 245.

OSELLA, A. M., and DUHAU, S. (1985), *Analysis of the Effect Produced by Lateral Inhomogeneities in the Mantle at Equatorial Latitude*, J. Geomag. Geoelectr. *37* (5), 531.

PĚČOVÁ J., MARTINEC, Z., and PEC, K. (1987), *Appreciation of Spherically Symmetric Models of Electrical Conductivity*, Pure Appl. Geophys. *125* (213), 291.

PRICE, A. T. (1950), *Electromagnetic Induction in a Semi-infinite Conductor with a Plane Boundary*, J. Mech. Appl. Math. *3*, 385.

PRICE, A. T., *Electromagnetic Induction within the Earth in Physics of Geomagnetic Phenomena* (Matsushita, S. and Campbell, C.) (Academic Press, New York 1967) p. 235.

RAMASWAMY, V., AGARWAL, A. K., and SINGH, B. P. (1985), *A Three-dimensional Numerical Model Study of Electromagnetic Induction around the Indian Peninsula and Sri Lanka Island*, Phys. Earth. Planet Inter. *39*, 52.

RITZ, M. (1983), *Use of Magnetotelluric Method for a Better Understanding of the West Craton Shield*, J. Geophys. Res. *88*, B12, 10625.

RITZ, M., and ROBINEAU, B. (1986), *Crustal and Upper Mantle Electrical Conductivity Structures in West Africa: Geodynamic Implications*, Tectonophysics *124*, 115.

ROBERTS, R. G. (1983), *Electromagnetic Evidence for Lateral Inhomogeneities within the Earth's Upper Mantle*, Phys. Earth. Planet. Inter. *33*, 198–212.

SIEBERT, M., and KERTZ, W. (1957), *Zur Zerlegung eines lokalen erdmagnetischen Feldes im äusseren und inneren Anteil*, Narch Akad. Wiss, Goettingen, Math-Phys. K15.

(Received February 15, 1990, revised/accepted June 25, 1990)

PAGEOPH, Vol. 134, No. 4 (1990)

0033–4553/90/040575–13$1.50 + 0.20/0

A Deep Geophysical Study in the Baikal Region

A. M. Popov[1]

Abstract—Experimental data show that in East Siberia resistivity curves, irrespective of their trends, are affected by galvanic (local) distortions. The preliminary step of the magnetotelluric data processing is to obtain a steady shape of resistivity curves reflecting a true deep section. For this purpose statistical averaging and different criteria of impedance rejecting were used. The available MTS curves were normalized by ρ_a level to the global magnetovariation curves. Two-dimensional modelling was performed from several sublatitudinal profiles crossing the Baikal rift zone. Three-dimensional models based on two-dimensional modelling and on induction vector distribution have been computed via programs of M. N. Yudin. Following other researchers, two conductive layers are distinguished: i) the mid- and low crustal and ii) the mantle one, with the layer surface uplifted from 100–110 km depth in the southern Baikal rift zone to 60–70 km northeastwards along the eastern Baikal coast. The top of this layer seems to correspond to the asthenospheric roof. The asthenosphere deepening in southern BRZ is likely to be related to a decrease in the asthenospheric bulge width and an increase in the rate of lithospheric thickening with mantle degasing. The origin and evolution of the Baikal rift is considered, proceeding from the model of passive rifting with regard to a long-existing lithospheric inhomogeneity between the Siberian platform and the Sayan-Baikal folded area.

Key words: Resistivity, ρ_a curves, MTS, impedance, heat flow, Baikal rift inhomogeneity, distortions.

Introduction

Numerous magnetotelluric soundings are carried out over the Baikal rift zone. Unfortunately, most resistivity curves are obtained within a narrow electromagnetic variation range reflecting mainly the middle lithosphere. Their shape and level show rather poor lateral correlation. Previous study has shown that most MTS curves are affected by galvanic distortions, as exemplified by the Irkutsk amphitheatre (POPOV, 1978; 1984, 1988a; POPOV and KUZMINYKH, 1988; POPOV *et al.*, 1987, 1988). A recent study of magnetotelluric data for the north-eastern Baikal rift zone reveals the same results (POPOV *et al.*, 1989). Due to the galvanic effects the detected layers are often misinterpreted now as crustal, now as mantle ones, though the effects are manifested within the same periods with the MTS sites located one near another. So it appears reasonable to prefer regional geoelectric sections reflecting a

---

[1] Institute of the Earth's Crust, USSR Academy of Science Siberian Branch, Irkutsk 664 033, U.S.S.R.

general regional setting rather than detailed but less reliable local ones. Therefore, a preliminary step in the magnetotelluric data analysis is important aiming to obtain a steady shape of the available resistivity curves which would reflect a true deep section.

In the present report I seek to show that the majority of the magnetotelluric soundings are distorted by the galvanic effects, consequently they are surely to be normalized by ρ_a level to the global magnetovariation curve.

Another segment of the paper contains the idea of heating of the sublithospheric material in stable regions. I believe that the activity of the rift zone results from a discharge of thermal energy accumulated beneath the adjacent craton. This idea substitutes for the "hot spot" hypothesis which seems hardly plausible to me.

Rifting mecahnisms proposed by other authors as well as the reported magnetotelluric study are reviewed in detail in POPOV (1988b) and POPOV (1989).

162 magnetotelluric soundings were considered for the Baikal rift zone. The effect of near-surface inhomogeneities on the impedances over the entire magnetotelluric variation range limiting the MTS was analysed. If these effects are stable within the whole resistivity curve we have a subparallel shifting of the curve related to the magnitude of distortion, i.e., the galvanic effects are manifested. Illustrating the above, Figures 1a,b show resistivity curves for short periods representing the upper section with the resistivity values successively for longer periods, at different distances from a two-dimensional inhomogeneity. The curves are plotted using the master curves of Dmitriev for H- and E-polarized fields (DMITRIEV et al., 1985). The parallel position of the resistivity curves relative to a distance from the inhomogeneity is known to be manifested with H-polarization. Such a position is illustrated by the relationship between the resistivity values for $\tilde{\lambda} = 4, 2$ representing the surface inhomogeneities and the resistivity for successively longer periods with $\rho\tilde{\lambda} = 4, 6, 10, 20, 40$ at points of different distances from the inhomogeneities (Figures 1a,b). These parallel plots show stable surface effects on the entire resistivity curve. E-polarization curves have slopes becoming progressively more shallow with longer periods. Here the two-dimensional near-surface effects attenuate as the period is increased (Figure 1b). The above regularities allow us to distinguish among practical magnetotelluric curves longitudinal (II) and transversal ($\perp$) ones, based on the analysis of impedances referred to the near-surface crust in correlation to fixed long-period variation impedances reflecting the deep section. If these parameters are closely related and slopes of regression plots ($\alpha_{1/2}\dots$, Figure 1) for all fixed periods are equal, it is safe to postulate that galvanic distortions are prevailing in the region under study. But if such relations are not valid or dip angles of regression plots decrease with increasing period, then we may deal with longitudinal curves.

In the inversion of practical curves, the parameters of $Z^{T=25}_{min,max}$ and $Z^{T=100,225,400,900}_{min,max}$ were correlated. Figures 1c,d contain the correlation results and the regression curves. With correlation coefficient 0.72–0.82 slopes at 95% con-

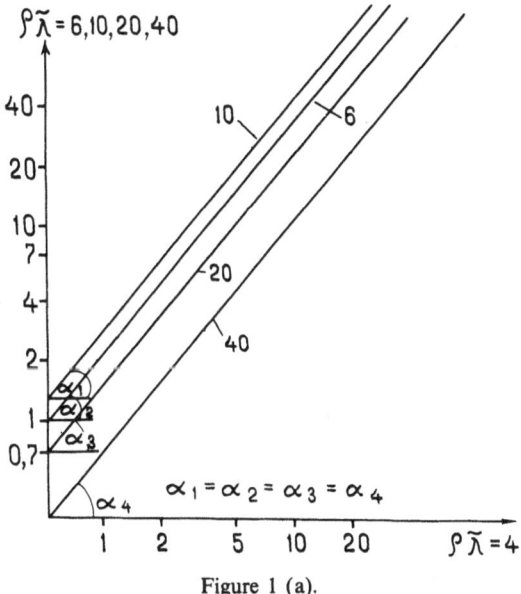

Figure 1 (a).

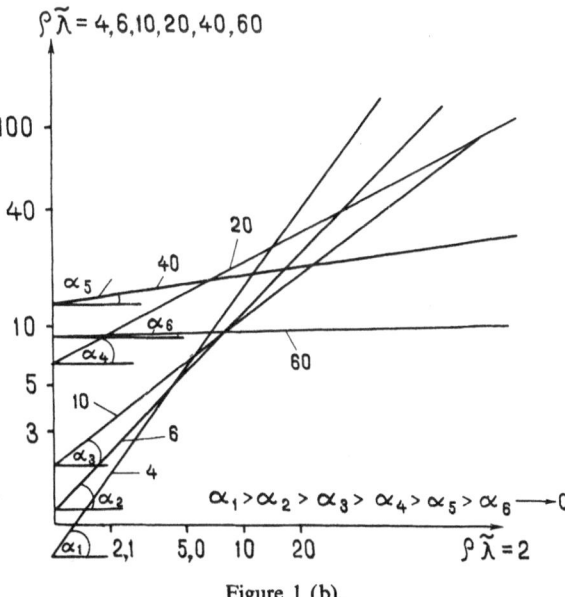

Figure 1 (b).

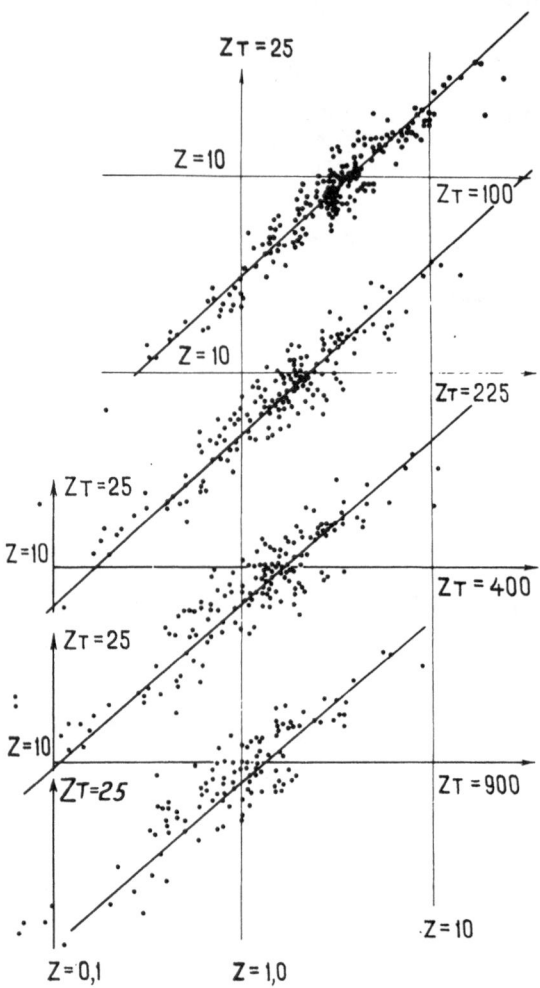

Figure 1 (c).

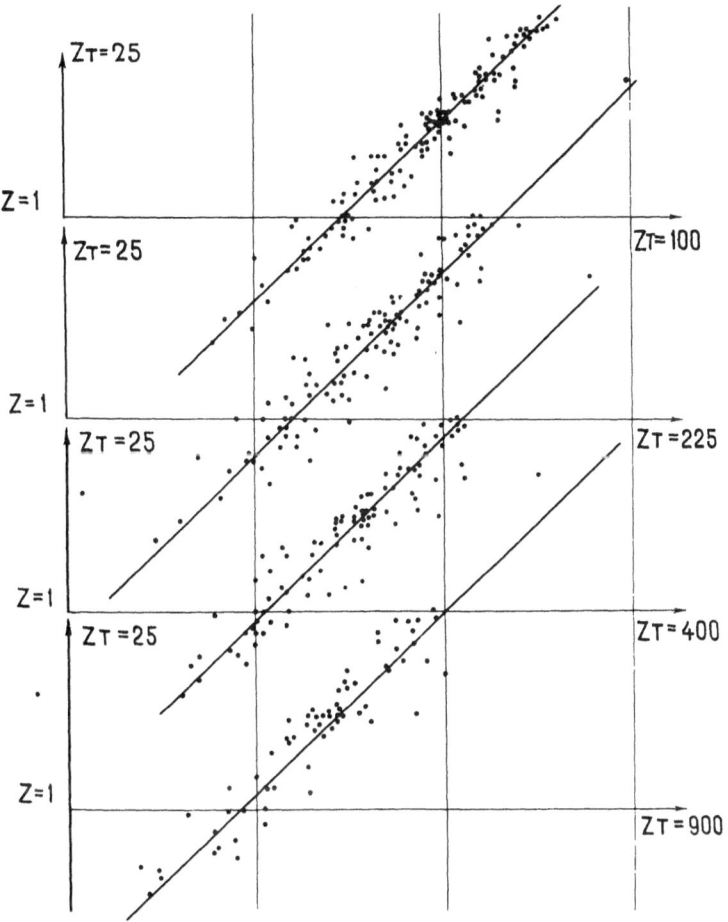

Figure 1

Apparent resistivity curves for the fixed lengths of electromagnetic wave $(\tilde{\lambda} = \lambda/h_1)$ equal to 4, 2 correlated to that for successively longer waves $(\tilde{\lambda} = 4, 6, 10, 20, 40)$ at a distance from a two-dimensional inhomogeneity, based on modelling data obtained by DMITRIEV et al. (1985). $a = H$-polarized field; $b = E$-polarized field; $c, d =$ correlation curves for maximum (c) and minimum (d) impedances with periods successively equal to $T = 100, 225, 400, 900$ s.

fidence level differ negligibly. The data obtained presume that, in the region under study, resistivity curves are subject to galvanic effects. Thus, here their normalization to global magnetovariation curves is possible.

It should be emphasized that nonconformity of resistivity curves (ρ_{max}, ρ_{min}) and the respective disturbed correlation relationship do not rule out galvanic effects. The shape of resistivity curves may be distorted in the sites of a complex geoelectric setting related to violated orthogonality of E- and H-components, driving this to scattered impedance values (POPOV and KUZMINYKH, 1988). The violated orthogonality between electric and magnetic fields must result from their inadequate behaviour: E-polarization strictly controlled by local inhomogeneities, and time dependent H-polarization not related to any local subsurface feature. Therefore, general subsurface inhomogeneities in the Baikal region practically rule out any possibility of obtaining longitudinal resistivity curves attributed to an electric field. In eastern Siberia the MT curves are affected mainly by galvanic distortions. Their interpretation reasonably requires statistical averaging or matching by ρ_a level to the global magnetovariation curve.

To obtain a steady shape of resistivity curves, different averaging methods and rejection of impedances (POPOV et al., 1987; POPOV et al., 1988) were used as well as the grouping of resistivity curves located within a 20–25 km radius along the profiles (Figure 2). Averaging radius has been determined proceeding from MT soundings' density and changes in the shape of resistivity (ρ_a) curves (the sought radius was limited by the value beyond which the shape of a ρ_a curve changed abruptly), as well as from the size of the area of effect ($r \sim \lambda/2\pi$) for the periods belonging to the S interval or to the beginning of the H interval. Averaged were mainly Z_{max} values since they are least distorted by some parameters (POPOV and KUZMINYKH, 1988; POPOV et al., 1987). The obtained averaged resistivity curves were normalized to long-period MT curves, and the latter were matched to the global MVS curve.

Averaging results for profiles extending along and across the Baikal rift zone are shown in Figure 3. One can see that averaged resistivity curves rather clearly reflect general regularities of the behaviour of conductive layers from the Siberian platform towards the Baikal rift zone and northeastwards along it. It is noteworthy that following the profiles in the above directions the two conductive layers extending into the crust (for $T = 100$ s) and into the mantle ($T = 200$ s) progressively transform into the one reflected by a long descending branch (to 700 s). Here the assumption may be valid that in these directions the asthenospheric layer is uplifting, and the crustal one is deepening. This leads to the thinning of the intermediate high resistive layer in the lithosphere. Therefore, the transformation of two-minimum curves into QHK (or KQHK) ones may be accounted for by the fact that magnetotelluric sounding is not sensitive to thin intermediate high resistive masses. Based on these ideas, mathematical modelling was performed. The electromagnetic field was calculated via programs of M. N. Yudin. We used a combined modelling. Along the profiles AB, AC and ED (Figure 2) the resistivity was modelled in 2D. Then three-dimensional models were constructed based on these results. The models were processed using an automatic choice of a grid with the size of its cell depending on the size of the inhomogeneities, distances and the thickness of the skin-layer (VANYAN et al., 1984). Modelling results are presented in Figure 3. Here the mantle conductive layer being evidently referred to the asthenosphere, is uplifting northeastwards within the rift zone from 110 to 60–70 km, and the crustal one is, on the contrary, deepening from 12–14 to 25–30 km. The resistivity of the mantle layer is estimated as 60 Ωm, and the crustal one amounts to 40–100 Ωm.

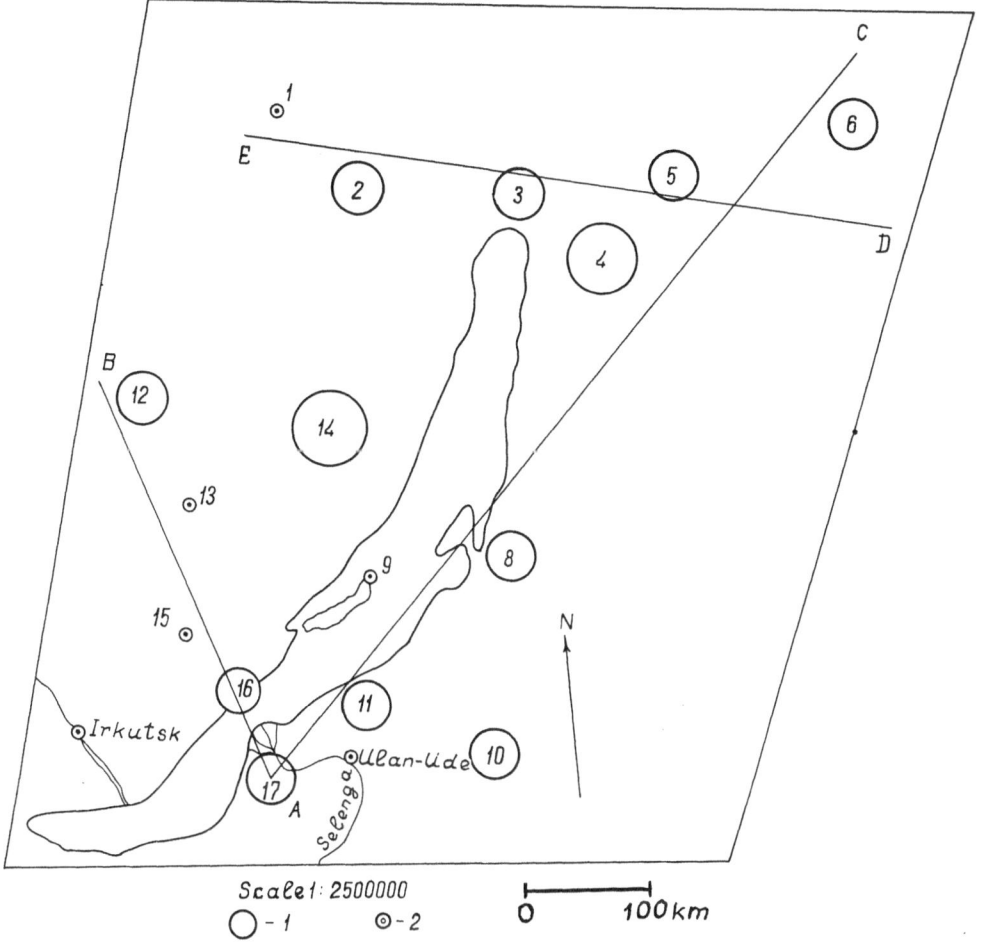

Figure 2
Areas covered by averagings (1); individual magneto-telluric soundings (2).

Such position of the asthenosphere is consistent with the petrological data on mantle xenoliths and with the heat flow data. These data indicate that partially melted material is lacking directly beneath the crust. It is stressed that only a few rocks of the mantle rock association show signs of partial melting. The data on amphibole and phlogopite-bearing lherzolites refute the idea of thermal homogeneity of highly heated (1200°C) anomalous mantle material beneath the crustal bottom, since the stability of pargasite amphibole at such depths does not exceed 1050°C (GREEN, 1973; KISELEV and POPOV, 1989).

Geothermometry and geobarometry data reveal the balance temperature values for the spinel peridotitic zone of the upper mantle ranging from 850 to 1200°C with a distinct maximum between 1050 and 1100°C (KISELEV and POPOV, 1989).

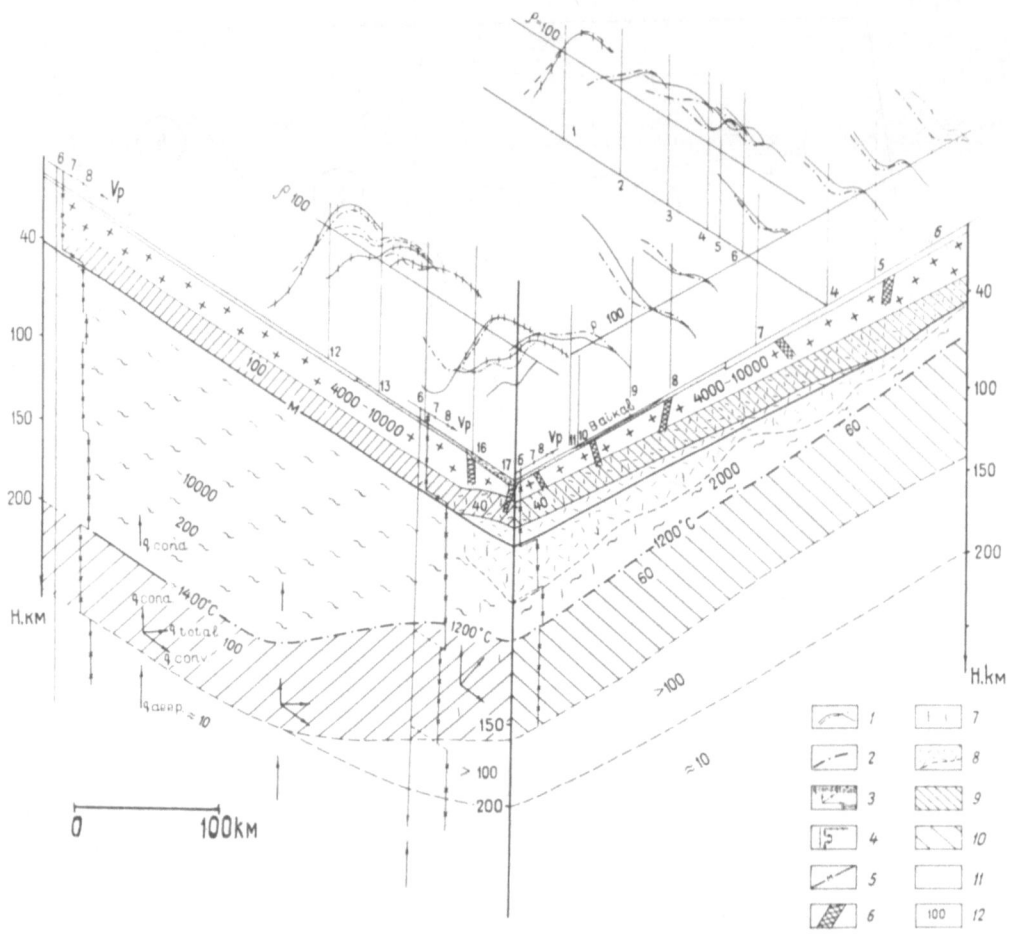

Figure 3

Block-diagram of geodynamics and deep structure of the contact zone between the Siberian platform
and the Sayan-Baikal folded area. 1 = experimental and model resistivity curves (their numbers
correspond to MTS grouping areas from Fig. 2); 2 = the asthenospheric top from seismological and
magnetotelluric data; 3 = heat flow vectors; q_d = deep heat flow, q_{cond} = conductive, q_{conv} = convective,
q_{total} = total ones; 4 = seismic velocities V_p versus depth plots, from DSS data; 5 = Moho discontinuity;
6 = zones of faulting; 7 = convective heat flow associated with active fluids; 8 = fracturing and ductile
stress zones, dashed line shows the bottom of the low-velocity lens, from DSS data; 9, 10 = crustal and
mantle high conductive layers; 11 = sediments; 12 = resistivity values in ohms.

Thus, we infer that mineral and phase composition of mantle rocks representing the
average depths of 60–65 km does not meet the notion of "anomalous mantle"
implying that the material contains no more than several per cents of the melted phase.
Therefore, it appears more reasonable to assume the existence of discrete masses of
the anomalous mantle rather than an integral continuous layer. Such masses fit the
conditions of maximum lithospheric extension (mainly beneath depressions) that

favours the mantle fluid transition from the asthenosphere, the metasomatism and heating of mantle material up to its partial melting. Thus, it follows that in the Baikal rift zone the convective heat transfer is prevailing. Intensive loss of thermal energy through weakened zones here is conditioned by its compensation from the depths, or *vice versa* the intensive heat evacuation results from the energy excess accumulated by the mantle. The position of the Baikal rift zone, at the contact between the Siberian platform and the Sayan-Baikal folded area, reflects the general regularity of concentric emplacement of continental rifts around cratons inheriting the ancient weakened zone of the lithosphere (DRAKE, 1972). Marginal suture is such a weakened contact zone. Tectonic activity associated with platform margins appears to be called forth by the release of energy accumulated beneath the stable zone.

Deep heat flow underneath the lithosphere involving radiation, excitation, conduction and convection components exceed by far its slower conductive evacuation through the lithosphere. This difference in heat flow may bring about heat accumulation beneath the lithospheric bottom. The amount of the accumulated heat would be controlled by melting and convection of sublithospheric material towards the weakened zone. This is evidently the reason of repeated tectonomagmatic activity of marginal stable areas (DRAKE, 1972; PAVLOV, 1978).

Figures 4a,b contain four models simulating heating processes, and the depths before the arising of convection towards weakened zones. They are respectively: (I) = before the Late Paleozoic activity, (II) = before the Early Mesozoic activity, (III) = before the Late Mesozoic, (IV) = before the Cenozoic one that involves rifting. The rate of subsidence of the asthenosphere decreases with depth since the mantle becomes depleted with basalt and fluid components.

Calculation results show that the heat flow in stable areas does not reflect the cycles of tectonomagmatic activity in marginal zones associated with convective heat flowing around the cratonic lithosphere. Thus stable and active areas appear to differ by the mode of heat transfer rather than by the extent of heating of the depths (cooler in stable areas and hotter in the active ones). The mode of heat transfer is conditioned by the lithospheric thickness and the Q-factor. Conductive heat transfer associates with the thick lithosphere of stable regions and the convective one prevails in the destructed permeable lithosphere of active zones.

The convective flow model yields considerably faster cratonisation without any constraints on lithospheric thickness (unlike the conductive cooling model). Lithospheric thickness is controlled not only by a cooling period but also by the size of a stable area, and structural anisotropy of the framing weakened zone, as well as by the global tectonic setting.

The origin and evolution of the Baikal rift can be explained proceeding from the model of passive rifting with regard to a long existing lithospheric inhomogeneity between the Siberian platform and the Sayan-Baikal folded area. This inhomogeneity, highly permeable, transformed the deep heat flow with a broken continuity at

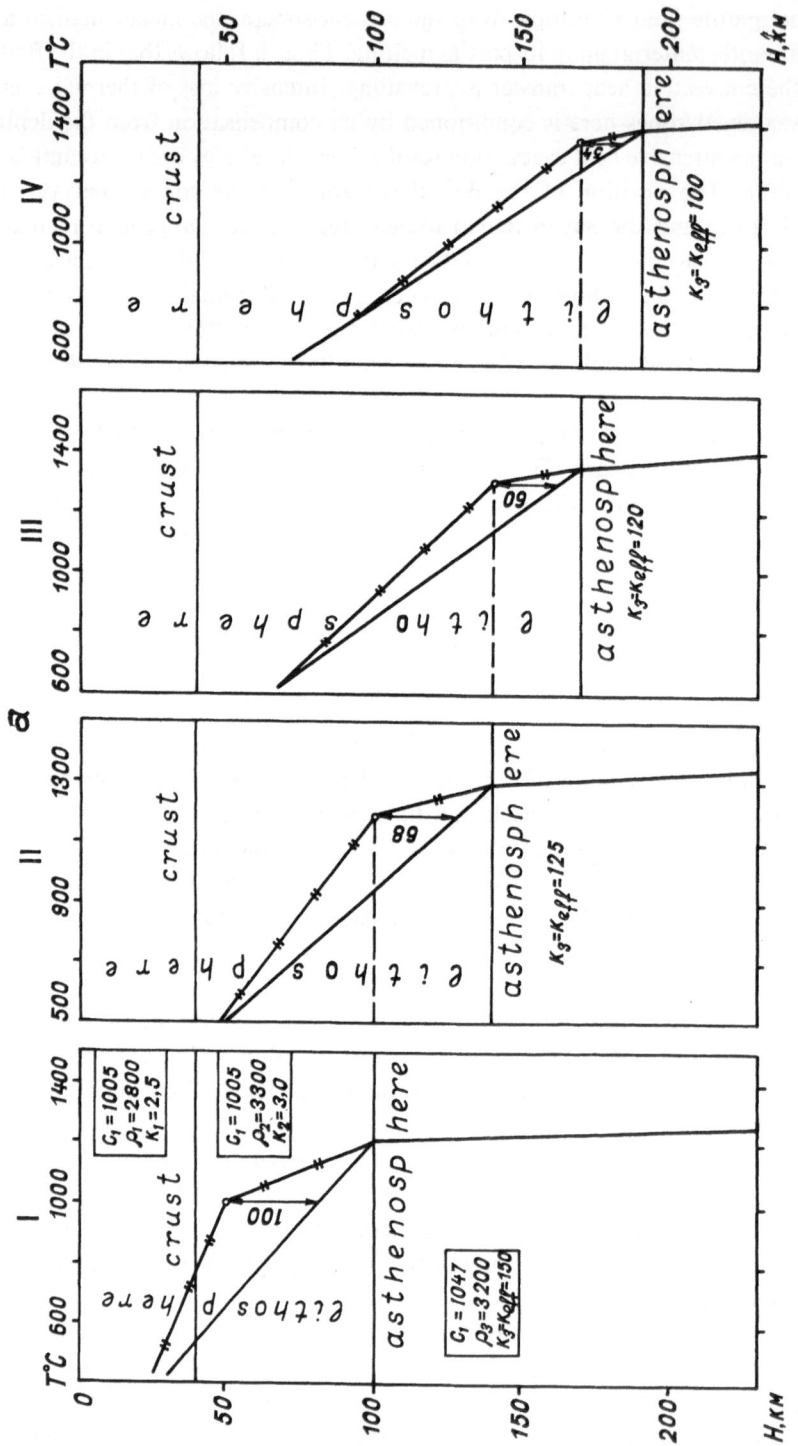

Figure 4(a).

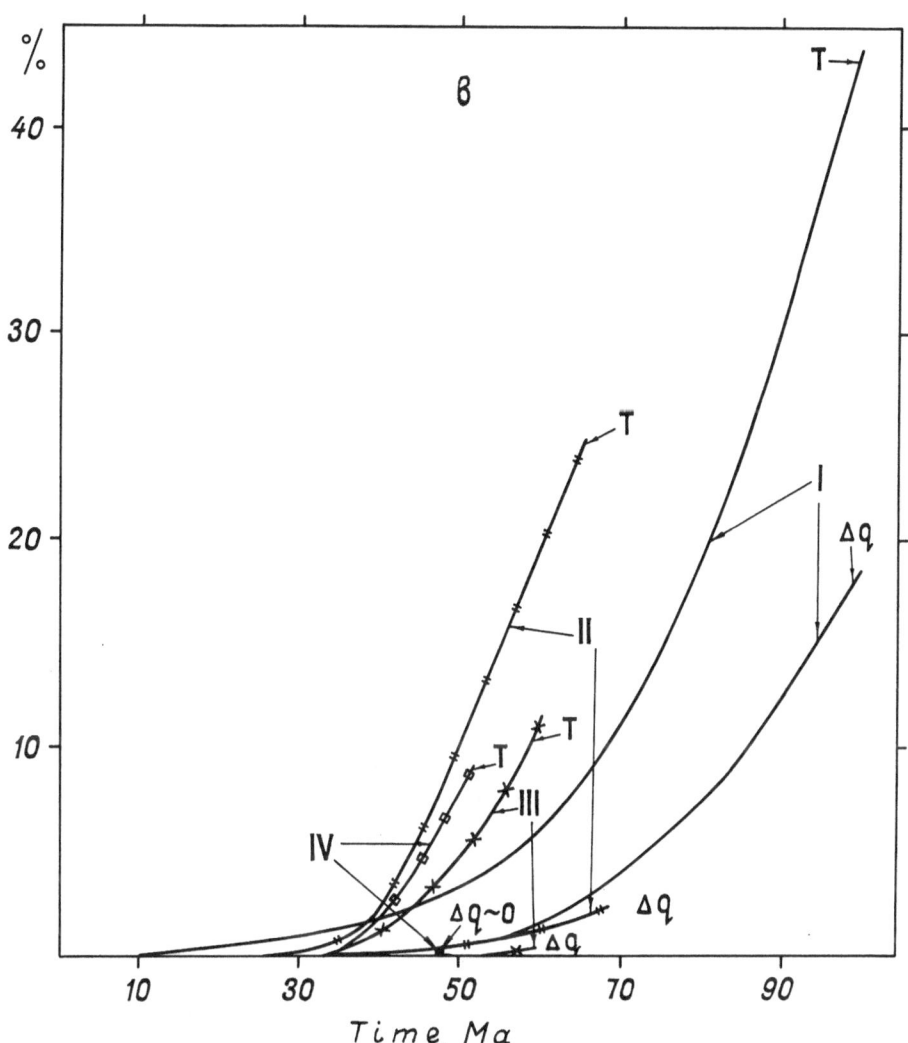

Figure 4

Mathematical modelling results.—sketch models of mantle heating in 4 periods of tectonomagmatic activity: I = before the Late Paleozoic, II = before the Early Mesozoic, III = before the Late Mesozoic, IV = before the Cenozoic, c = heat capacity; ρ = density; K = heat conductivity. Solid line designates initial temperature distribution; cross-hatched line shows mantle heating in respective depths: I = 50 km, II = 100 km, III = 140 km, IV = 170 km. Vertical arrow shows heating range; number near the arrow denotes heating duration in Ma. See text for details. b—curves of temperature variations (in per cent of initial values) with heating associated with the asthenospheric formation depths (100, 140, 170 and 190 km), together with curves of surface heat flow variations. Roman numerals correspond to periods from Fig. 4a.

the bottom of subcratonic lithosphere, into conductive and convective components. The convective heat flow, oriented sublaterally towards the folded area, kept up the weakened lithosphere along the boundary of the above major structures. Stress redistribution within the Eurasian plate, induced by the Eocene collision, must have called forth the reactivation of the weakened lithospheric zone that has determined the subsequent formation of the Baikal rift.

Acknowledgements

The author is grateful to M. N. Yudin for kindly offering computer programs, and to Mrs. T. Perepelova for her assistance in preparing the English version of the paper.

REFERENCES

DMITRIEV, V. I., BERDICHEVSKY, M. N., and KHOKHOTUSHKIN, G. L., *Master Curves for MT Soundings in Inhomogeneous Media. Part 4* (MGU, Moscow 1985) p. 100 (in Russian).

DRAKE, C. L., *Continental margins*, In *Crust and Upper Mantle* (Mir, Moscow, Russian translation 1972) pp. 473–480 (in Russian).

GREEN, G. N. (1973), *Experimental Melting Studies on Model Upper Mantle Composition at High Pressure under Both Water-saturated and Water-undersaturated Conditions*, Earth. Planet. Sci. Lett. *19* (1), 37–53.

KISELEV, A. I., and POPOV, A. M. (1989), *Structure and State of Depths beneath the Baikal Rift from Petrological and Geophysical Data*, Doklady Akad. Nauk SSR. *306* (4), 938–941 (in Russian).

MOLNAR, P., and TAPPONNIER, P. (1975), *Cenozoic Tectonics of Asia: Effect of Continental Collision*, Science *189* (4201), 419–426.

PAVLOV, S. F. (ed.), *Relations between Ancient and Cenozoic Structures in the Baikal Rift Zone* (Nauka, Moscow 1978) p. 125 (in Russian).

POPOV, A. M., *Deep MT soundings in the Baikal region*, In *Seismicity and Deep Structure of the Baikal Region* (Nauka, Novosibirsk 1978) pp. 94–101 (in Russian).

POPOV, A. M., *On The effect of local surface inhomogeneities on MT field structure*, In *Geomagnetic, Aeronomic and Sun Physics Study* (Nauka, Moscow 1984) pp. 190–196 (in Russian).

POPOV, A. M. (1988a), *Effect of Subsurface Inhomogeneities on MTS Data*, Izvestia Akad. Nauk SSR, Fizika Zemli *3*, 87–91 (in Russian).

POPOV, A. M. (1988b), *Deep geoelectric structure of the Baikal region and its tectonic interpretation (from magnetotelluric data along the profile Zhigalovo-Red Chikoy)*, In *Asthenosphere from Complex Geophysical Data* (Naukova Dumka, Kiev 1988) pp. 201–207 (in Russian).

POPOV, A. M. (1989), *Results of Deep Magnetotelluric Soundings in the Light of Other Geophysical Studies in the Baikal Region*, Izvestia Akad. Nauk SSSR, Fizika Zemli *8*, 31–37 (in Russian).

POPOV, A. M., and KUZMINYKH, Yu. V. (1988), *On observation and interpretation methods for MT soundings in the Baikal Region*, In *Geomagnetic Study 31* (Nauka, Moscow 1988) pp. 39–49 (in Russian).

POPOV, A. M., KUZMINYKH, Yu. V., and BADUEV, A. B. (1989), *Experimental Study of Local Inhomogeneities in the MT Field*, Geol. Geophys. *9*, 117–127 (in Russian).

POPOV, A. M., POTAPOV, A. S., KUZMINYKH, Yu. V. *et al.*, *Standard MT soundings for the Southern East Siberia*, In *Geomagnetic, Aeronomic and Sun Physics Study*, Vol. 78 (Nauka, Moscow 1987), 125–132 (in Russian).

POPOV, A. M., POTAPOV, A. S., KUZMINYKH, Yu. V. *et al.* (1988), *Long-period MT Soundings in the Baikal Region*, Izvestia Akad. Nauk SSSR, Fizika Zemli *11*, 77–81 (in Russian).

VANYAN, L. L., DEBABOV, A. S., and YUDIN, M. N., *Interpretation of MT Data in Inhomogeneous Media* (Nedra, Moscow 1984) p. 197 (in Russian).

(Received February 15, 1990, accepted August 15, 1990)

PAGEOPH, Vol. 134, No. 4 (1990)

0033–4553/90/040589–10$1.50 + 0.20/0

A New Telluric KCl Probe Using Filloux's AgAgCl Electrode

Andreas Junge[1]

Abstract — The measurement of the telluric field's long-time variations requires stable instruments in the period range above 1 day. Obviously, most problems arise from drifting voltages between the telluric probes. Good results have been achieved using a three chamber Hempfling KCl probe together with Filloux's AgAgCl electrode. However, a one chamber probe of 0.6 m length and 0.06 m diameter filled with saturated KCl solution may be sufficient for some applications and additionally allows permanent control of the electrolyte's salt concentration. In a field test the telluric field at a single site was simultaneously observed using one and three chamber probes separated by 25 and 55 m resp. For periods shorter than 1 hour the noise level of the electrodes was found to be less than 1 $(mV)^2/Hz$ whereas in the period range between 1 hour and 1 day it increases to 100 $(mV)^2/Hz$. Consequently, surveys investigating long periods of the telluric field can be carried out using small electrode separations of a few tens of meters. Furthermore, the stability of this probe negates the need to interrupt the time series for servicing of the probe. However, the one chamber probe is somewhat inferior to the three chamber probe with regard to temperature dependence at long periods.

Key words: Telluric field, KCl probe, Long-time variations, electrode noise.

Introduction

The measurement of the time varying telluric field generally requires two buried probes and the recording device which includes electronic filtering and amplification of the potential difference between the probes. In the long period range above one hour the noise in the measurement is due to the noise and drift properties of the probes. Consequently, the properties of the probes used in Göttingen were examined and where necessary improved.

Principle of Measurement and Some Previous Results

Petiau and Dupis (1980) give an overview of different electrode types. For periods longer than 0.1 s electrodes of the type described below prove to be best. These electrodes consist of metal which is in equilibrium with a not readily soluble salt of its ions. The anions of the combination are contained in the soluble salt in

---

[1] Institut für Geophysik, Postfach 2341, D-3400 Göttingen, FRG.

the electrolyte. Examples of such electrodes are silver–silverchloride electrodes (Ag/AgCl, KCl(c) + aq//), calomel electrodes (Pt/Hg/Hg$_2$Cl$_2$, KCl(c) + aq//), lead–leadchloride electrodes (Pb/PbCl$_2$, KCl(c) + aq//).

PETIAU and DUPIS (1980) prefer leadchloride electrodes based on their laboratory experiments. HEMPFLING (1977), however, recorded the Sq variation of the telluric field with AgCl electrodes separated by only 200 m. The signal amplitude was of the order of 1 mV. HEMPFLING used a 3-chamber probe filled with a KCl solution of which the concentration increased from the outer to the inner chamber (10%, 20%, saturated). FILLOUX'S (1967) AgCl-electrode was placed in the inner chamber. The probe was buried at about half a meter depth in the ground and could not be accessed during the field survey without interruption of the time series.

Since the electrolyte's concentration and hence the selfpotentials of the electrodes continuously change during each survey, a severe voltage drift generally results. This suggests that probe construction needs improvement.

The Electrochemistry of the AgAgCl-electrode

Following Nernst's equation the selfpotential ε_0 of the AgAgCl-electrode is

$$\varepsilon_0 = E_{0,\,\mathrm{Ag}} + \frac{RT}{F} \ln P_{\mathrm{AgCl}} - \frac{RT}{F} \ln c_{\mathrm{Cl}-} \tag{1}$$

$E_{0,\,\mathrm{Ag}} = 779$ mV is the normal potential of silver, $RT/F = 25.6$ mV is a scaling factor containing the universal gas constant R, the absolute temperature T and the Faraday constant F, $P_{\mathrm{AgCl}} = 1.8 \cdot 10^{-10}$ is the precipitation value of silverchloride, $c_{\mathrm{Cl}-}$ is the concentration of the chloride ions (VETTER, 1961). All the given numbers are valid at a temperature of 25°C. At the contact zone of two different electrolytes, e.g., probe/soil, a diffusion potential ε_D^{12} will occur due to different ion velocities. The diffusion potential is described by the Henderson equation

$$\varepsilon_D^{12} = \frac{\sum\limits_j \frac{u_j}{z_j}(c_{j,\,2} - c_{j,\,1})}{\sum\limits_j u_j (c_{j,\,2} - c_{j,\,1})} \frac{RT}{F} \ln \frac{\sum\limits_j u_j c_{j,\,2}}{\sum\limits_j u_j c_{j,\,1}} \tag{2}$$

where z_j is the valency, u_j the velocity, and $c_{j,\,1} c_{j,\,2}$ the concentration j of ion 1 and 2, respectively. Using a saturated KCl solution the term c_{KCl} will dominate the other terms by orders of magnitude, so that eq. (2) simplifies

$$\varepsilon_D^{12} \approx \frac{(u_{\mathrm{K}+} - u_{\mathrm{Cl}-})}{(u_{\mathrm{K}+} + u_{\mathrm{Cl}-})} \frac{RT}{F} \ln \frac{(u_{\mathrm{K}+} + u_{\mathrm{Cl}-}) c_{\mathrm{KCl}}}{\sum\limits_j u_j c_{j,\,1}}. \tag{3}$$

The potential difference U_0 between two probes buried in a homogeneous ground will be, with indices 1 and 3 for the probes and 2 for the ground

$$U_0 = \varepsilon_0^3 - \varepsilon_0^1 + \varepsilon_D^{32} - \varepsilon_D^{12} \approx \left[\frac{(u_{K^+} - u_{Cl^-})}{(u_{K^+} + u_{Cl^-})} - 1 \right] \frac{RT}{F} \ln \left(\frac{c_3}{c_1} \right). \qquad (4)$$

The value of the term in the square brackets is -1.02 for KCl and -1.20 for NaCl and CuSO$_4$; hence the sensitivity of U_0 towards changes in temperature and concentration is diminished using a saturated KCl solution. Futhermore, the saturated solution produces a KCl sediment which guarantees a long-time stability of the selfpotential, thus making the one chamber probe competitive with the three chamber probe. The importance of a constant salt solution is seen immediately, for a change in concentration of 1% will alter U_0 by 0.25 mV.

Structure of the One Chamber Probe

Figure 1 shows the structure of the probe. It consists of two transparent PVC cylinders which are screwed together. Four ceramic diaphragms are set into the sides of the lower cylinder providing electric contact between the interior of the probe and the ground. The lower cylinder contains Filloux's AgCl-electrode, the oversaturated KCl solution and the KCl sediment which should reach the lower edges of the diaphragms. The socket of the AgCl-electrode is glued into the bottom of the upper cylinder which contains two pipes filled with saturated KCl solution. Each pipe is closed at the top with a plastic screw and an airtight rubber sealing which serves to control the condition of the probe and to refill the lower cylinder, if necessary. All parts are glued with Tangit, except the ceramic diaphragms where Uhu Plus was used. The AgCl-electrode is connected to the electronic device by insulated copper wire of 0.3 mm diameter and 20 m length including a tension relief.

Fieldtest

Telluric and magnetic field variations were recorded simultaneously at Breitnau in the Black Forest during the summer of 1987. For the telluric field measurement one chamber and three chamber probes were arranged according to Figure 2. Configuration *a* refers to the one chamber probes and configuration *b* the three chamber probes. The probes of each configuration were connected to a separate recording unit with a sampling rate of 30 s and a 240 s lowpass filter. Figure 3a shows an 8-hour interval of simultaneous magnetic and telluric field records (E_{N1} and E_{E1} for the one chamber, E_{N3} and E_{E3} for the three chamber probes). There is a good correlation between the magnetic and telluric variations and an excellent agreement between parallel telluric field lines.

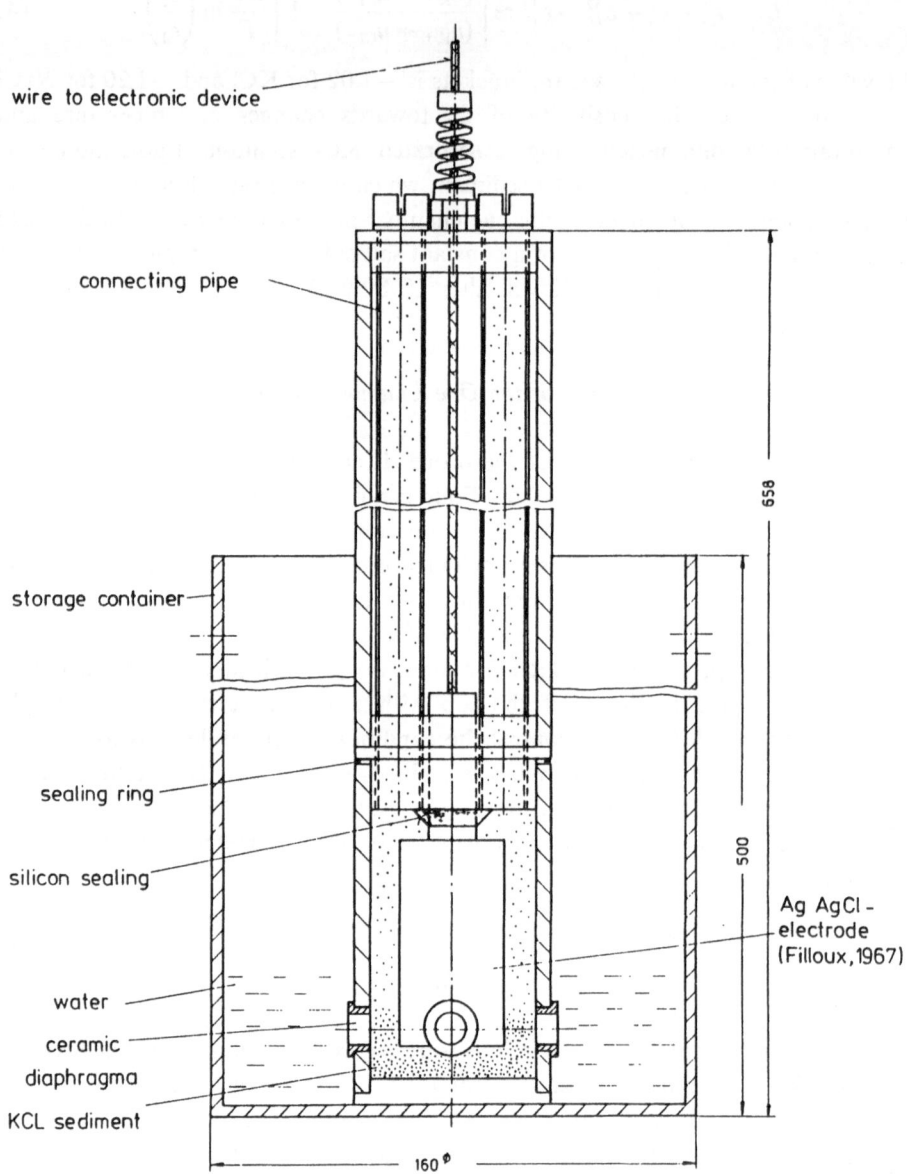

Figure 1
An accurate scale longitudinal section of the one chamber probe.

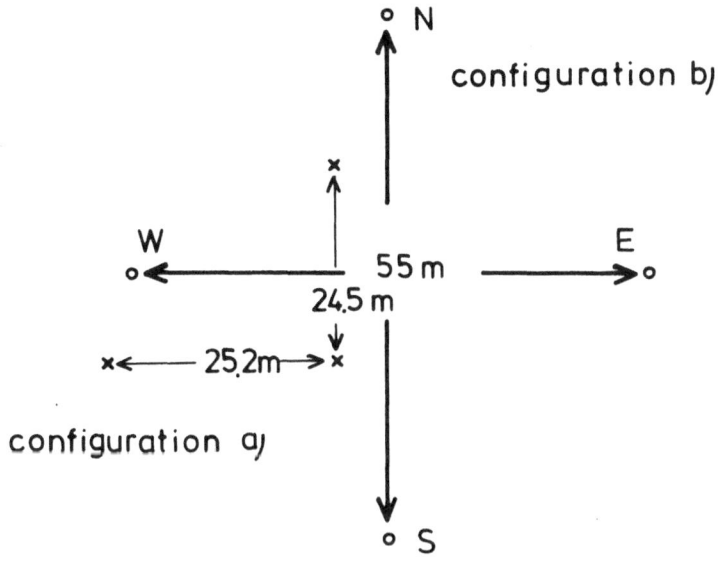

Figure 2
Probe configuration at BRE (Breitnau/Black Forest). a) One chamber probe (×) using "L-configura-
tion" (Common southern and eastern probe). b) Three chamber probe (○).

Figure 3b demonstrates power spectra which belong to the time series of Figure 3a for the frequency range of 0.125–10 cph. Both telluric field spectra are characterized by a decay of 4 decades towards the high frequency end. The spectrum of E_{N13} (the difference between E_{N1} and E_{N2}), however, falls off about one decade from 0.1 to 0.01 $(mV/km)^2 h$ and lies almost 3 decades below the other two spectra.

It follows with an underlaying 55 m probe separation that the noise energy of the electrodes is less than 1 $(mV)^2/Hz$ for the frequency range above 1 cph. Figures 4a, b present similarly to Figures 3a, b a ten-days interval of low-pass filtered (cut-off period 1 hour) values of the magnetic and telluric field. In the magnetic field the daily variations are clearly seen, but in the telluric field short periodic variations dominate. The fast variations of both telluric records agree very well which is demonstrated by the smooth curve of E_{N13}. Nevertheless there is a remarkable correspondence of this curve with the air temperature shown on top, not only for the daily variations but also for a long-periodic variation of about 5 days length. The comparison between the records of E_{N1} and E_{N3} demonstrates that the variations correlated with temperature obviously result from the one chamber probes. This assumption is confirmed by the power spectra in Figure 4b: For frequencies above 1 cpd the spectrum of E_{N13} lies more than 1 decade below the spectra of E_{N1} and E_{N3}, in the low frequency range it increases similarly to the temperature spectrum with peaks at 0.2, 1 and 2 cpd. This increase is also found in the spectrum of E_{N1} but not in that of E_{N3}.

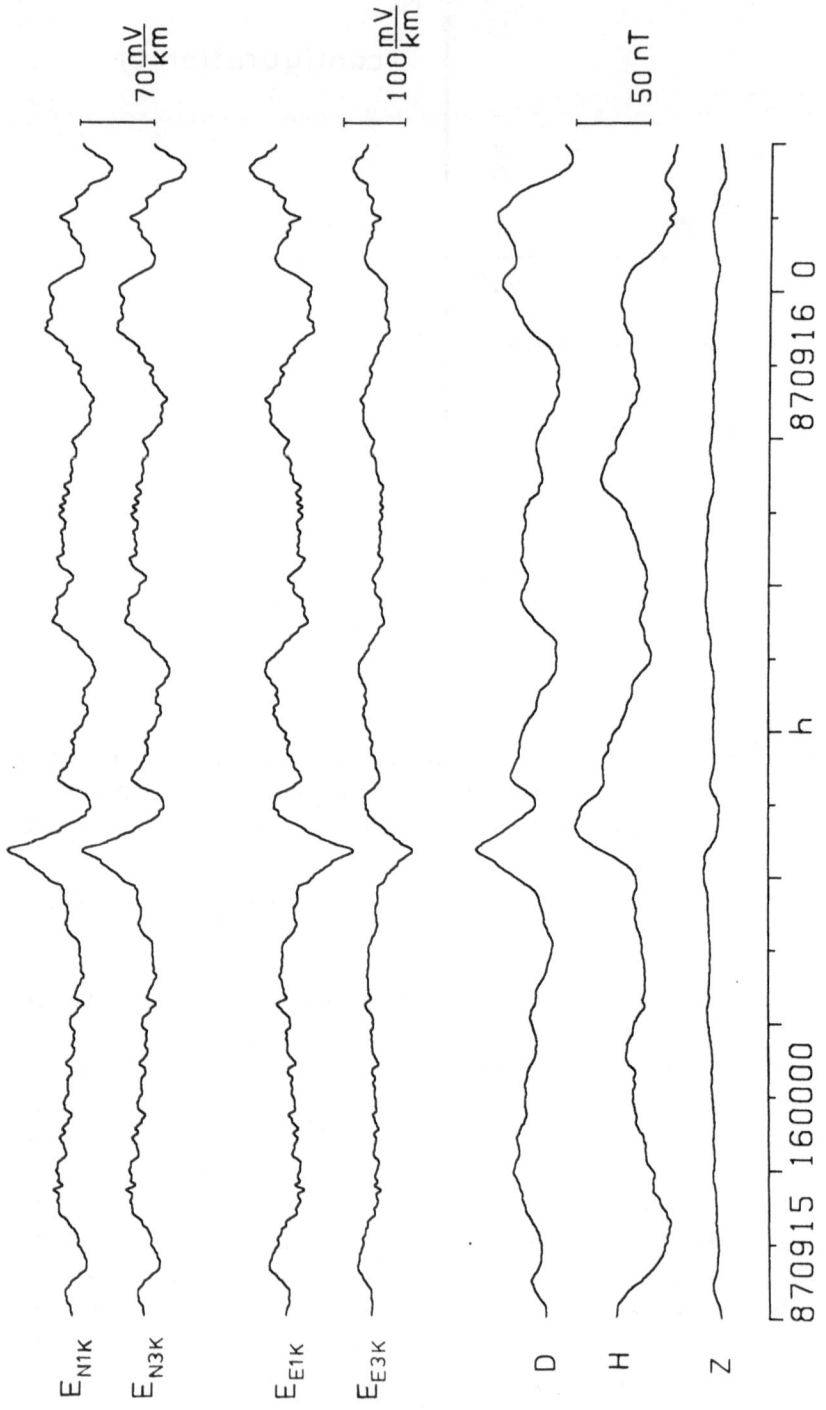

Figure 3 (a)

Record of 8 hours. H, D, Z components of the magnetic field, E_N, E_E telluric field in North and East direction, indices 1 and 3 for one chamber and three chamber probes.

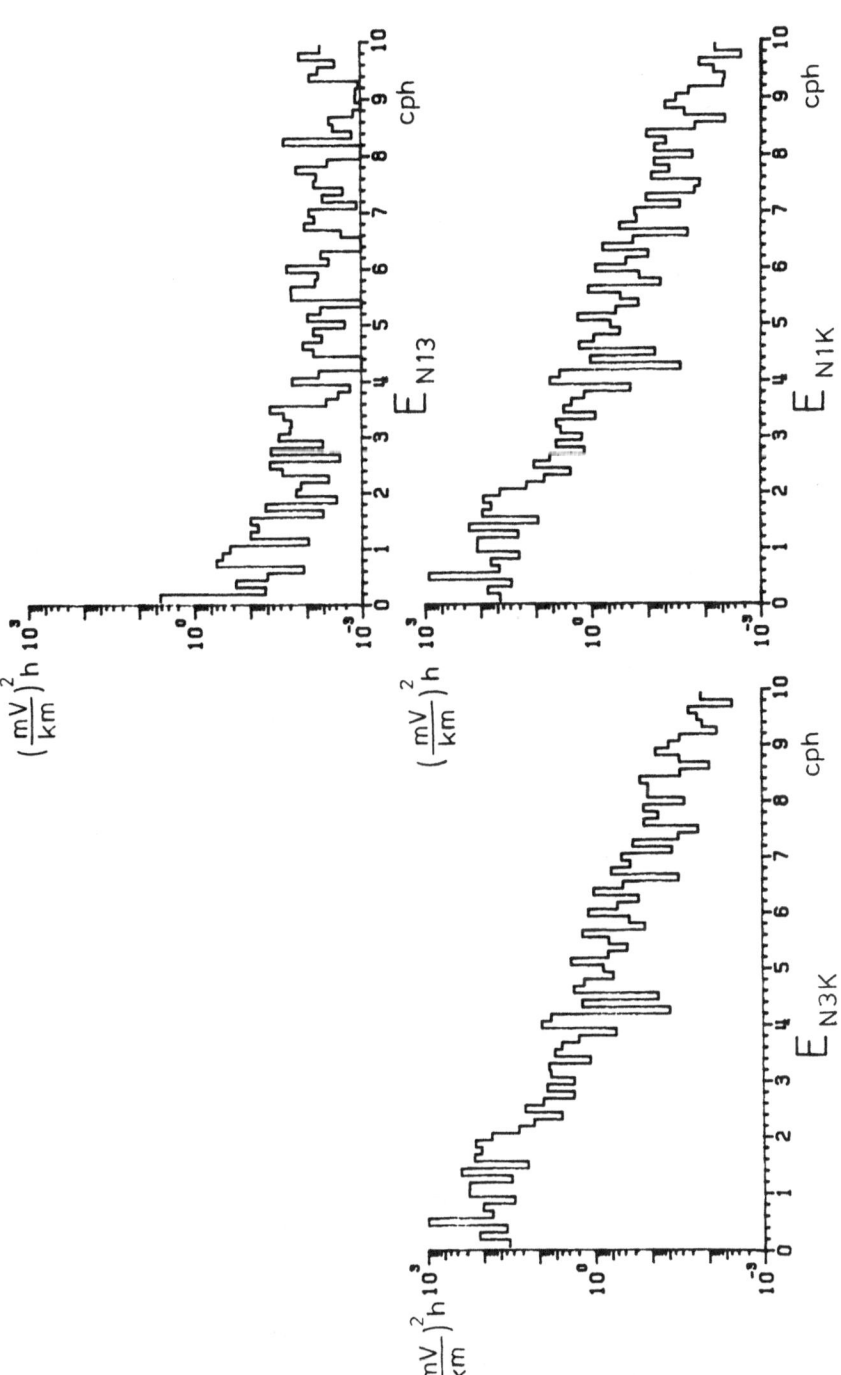

Figure 3 (b)

Power spectra of the time series E_{N1}, E_{N3} and E_{N13} in Fig. 3a for the frequency range 0.125–10 cph with a frequency spacing of 0.125 cph. The "noise spectrum" $E_{N13} = E_{N1} - E_{N3}$ is almost 3 decades below the "signal spectra"!

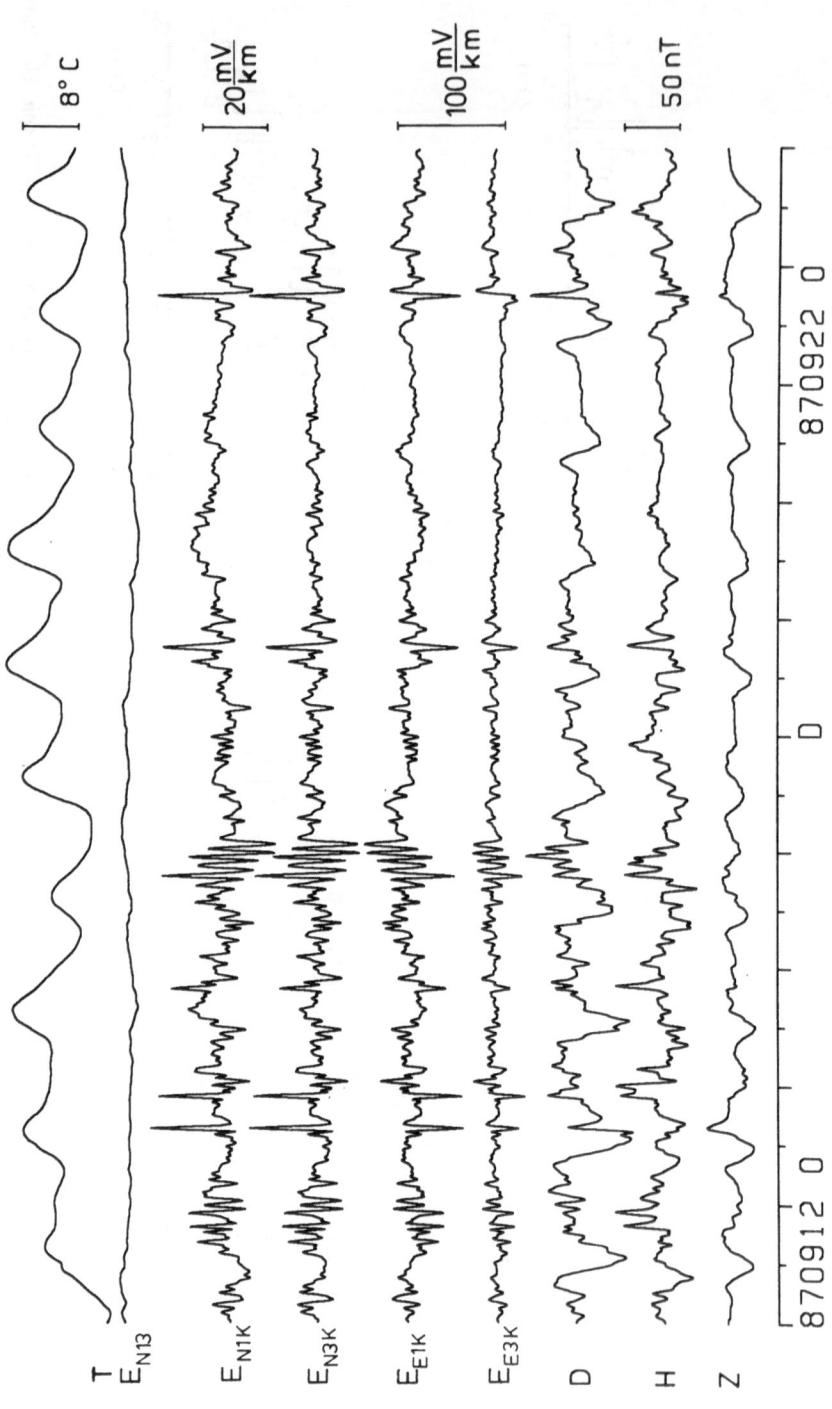

Figure 4 (a)

Low-pass filtered values of a 10-days registration (see also description of Fig. 3a). T is the air temperature. Remarkably the correspondence of T, E_{N13} and E_{N1} which hints at a temperature dependency of the one chamber probe.

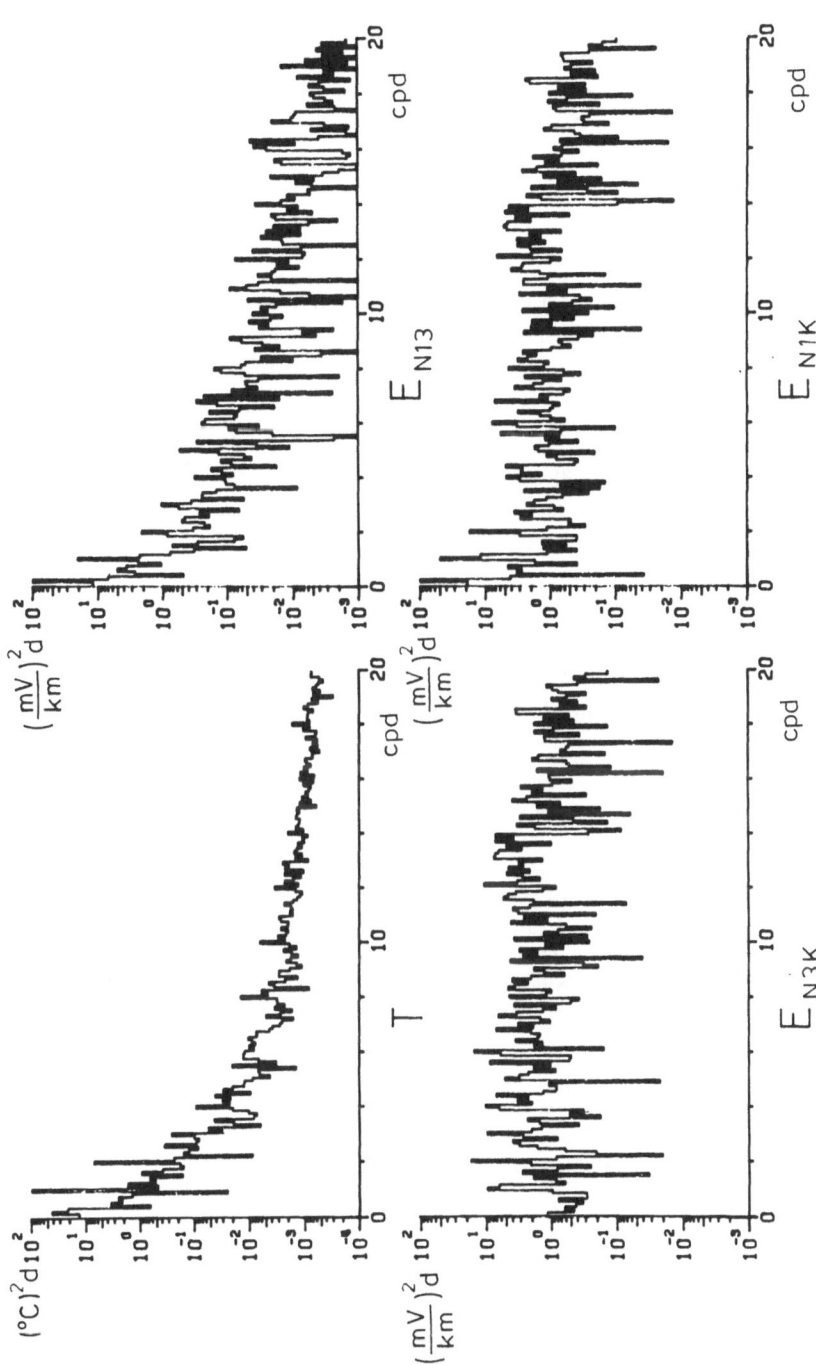

Figure 4 (b)

Power spectra of E_{N1}, E_{N3}, E_{N13} and T for Fig. 4a for the frequency range 0.1–10 cpd with a frequency spacing of 0.1 cpd. Remarkably is the similar decay of the spectra of T and E_{N13} between 0.1 and 3 cpd with significant lines of 0.2, 1 and 2 cpd.

The temperature behaviour of the one chamber probe obviously results from contact to the earth's surface, which is the subject of present investigations. However, above 1 cpd the spectrum of E_{N13} falls off from 0.5 $(mV/km)^2$ d to 0.005 $(mV/km)^2$d at 20 cpd. It follows that the noise level of the electrodes descends from 100 $(mV)^2/Hz$ at 1 cpd to 1 $(mV)^2/Hz$ at 20 cpd.

Conclusion

In the frequency range above 1 cph the noise behaviour of the probes agrees with the results of PETIAU and DUPIS. For lower frequencies, however, the probes or rather the AgAgCl electrodes constructed by FILLOUX show a noise level which is by orders of magnitude below that found by PETIAU and DUPIS. This is even more remarkable as PETIAU and DUPIS assume that the voltage variations of their AgAgCl electrodes result from aging material during an experiment lasting two years, but FILLOUX's electrodes used in the Black Forest experiment were at least several years old. NEURIEDER (1984) yields similar results with his investigations of Hempfling's three chamber probes. Although the one chamber probe is inferior to the three chamber probe with regard to sensitivity at temperature change, it has some operational advantages, particularly with respect to long-term stability. Furthermore, its basic stability might permit a rather short probe separation of a few tens of meters, thus reducing the possibility of damage to the connecting wire.

REFERENCES

FILLOUX, J., *Instrumentation and experimental methods for oceanic studies*, In *Geomagnetism* Vol. 1 (ed. Jacobs, J. A.), (Academic Press, London 1987).
HEMPFLING, R. (1977), *Beobachtung und Auswertung tagesperiodischer Variationen des erdelektrischen Feldes in der Umgebung von Göttingen*, Diss. math. nat. Fak. Univ. Göttingen.
NEURIEDER, P. (1984), *Die elektrische Leitfähigkeit des Oberen Mantels unter Mitteleuropa, abgeleitet aus langperiodisch variierenden, natürlichen elektromagnetischen Wechselfeldern*, Diss. Fak. f. Geowiss. Univ. München.
PETIAU, G., and DUPIS, A. (1980), *Noise, Temperature Coefficient, and Long-time Stability of Electrodes for Telluric Observations*, Geophys. Prosp. *28*, 792–804.
VETTER, K. J., *Elektrochemische Kinetik* (Springer Verlag 1961).

(Received January 31, 1990, revised/accepted August 21, 1990)